S0-ADX-022

Global Environmental Negotiations and US Interests

Global Environmental Negotiations and US Interests

Deborah Saunders Davenport

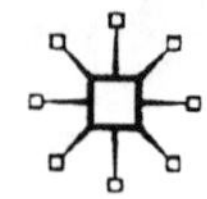

GLOBAL ENVIRONMENTAL NEGOTIATIONS AND US INTERESTS

First published in 2006 by
PALGRAVE MACMILLAN™
175 Fifth Avenue, New York, N.Y. 10010 and
Houndmills, Basingstoke, Hampshire, England RG21 6XS
Companies and representatives throughout the world.

PALGRAVE MACMILLAN is the global academic imprint of the Palgrave Macmillan division of St. Martin's Press, LLC and of Palgrave Macmillan Ltd. Macmillan® is a registered trademark in the United States, United Kingdom and other countries. Palgrave is a registered trademark in the European Union and other countries.

ISBN-13: 978–1–4039–7021–3
ISBN-10: 1–4039–7021–1

Library of Congress Cataloging-in-Publication Data

Davenport, Deborah Saunders.
Global environmental negotiations and US interests / Deborah Saunders Davenport.
p. cm.
Includes bibliographical references and index.
ISBN 1–4039–7021–1
1. Environmental management—International cooperation. 2. Environmental policy—International cooperation. 3. Climatic changes—International cooperation. I.Title.

GE300.D38 2006
363.73'526-dc22 2006041592

A catalogue record for this book is available from the British Library.

Design by Newgen Imaging Systems (P) Ltd., Chennai, India.

First edition: October 2006

10 9 8 7 6 5 4 3 2 1

Printed in the United States of America.

In remembrance of
Frederick S. Davenport, my father,
and Michael Nicholson, my first IR professor

Contents

Tables and Figures

Tables

Figures

Acknowledgments

For their never-ending support and encouragement in bringing this idea to fruition, I would like to thank Walter Mattli and Suzanne Werner. My thanks likewise to David R. Davis and Richard Rubinson for all their encouraging comments and advice. I would also like to thank Hannah Britton, Elisabeth Corell, Radoslav Dimitrov, Lars Gulbrandsen, David Humphreys, Ian Symons, Jane Winzer, and Amanda E. Wooden, as well as several anonymous reviewers and participants at the Nordic Forest Researchers Workshop in Stockholm, Sweden, September 2004 and the International Studies Association Annual Conventions in 2003, 2004, and 2005, for comments on earlier versions of some of the chapters presented here.

I also want to express my tremendous gratitude for the support given to me by several institutions with which I have been associated. I owe a great debt to many individuals who have been connected to Emory University and the Carter Presidential Center, including my friend Jane Winzer, Gordon Streeb, the late Andy Agle, the late Bebbi Johnson, Bill Mankin, and Robert Pastor, whose ability to bridge the worlds of academia and policy-making brought about my own efforts in this regard. During my time at Mississippi State University, Dave Breaux and Phil Oldham gave me the gift of time that allowed me to work on this project, for which I am grateful, and I was assisted ably by LaKesha Perry, Niles Puckett, Cortney Harris, and Christina Lee; my wholehearted thanks to all of them. I must direct special thanks to the International Institute for Sustainable Development's Reporting Services for giving me the opportunity to witness firsthand the negotiations that inspired this project as a reporter for the *Earth Negotiations Bulletin*. My heartfelt thanks to my colleagues in all these institutions, my students at Emory and MSU, and to all the people involved in the international negotiations discussed here whose insights have given me food for thought.

My deepest thanks go to my mother, Iris Davenport, for getting me all the way through this, in more ways than one, and to the other members of my family who have given me their constant support: Melanie Davenport, Virginia Davenport, Steve Burnley, Marie Burnley, Robert Yother, my husband, Rayman Mohamed, and my most exceptional helper and biggest buddy, Adam Mohamed-Davenport. Any errors are, of course, all my own.

CHAPTER 1

An Introduction to Three Global Environmental Issues

Why are some global environmental issues covered by binding, enforceable commitments for their amelioration while others are not? Why is there a binding international treaty system that is predicted to bring about full recovery of the ozone layer within the next 50 years, while there is no global treaty covering deforestation despite an "alarmingly high rate" of deforestation?[1] Finally, why, despite the existence of a binding international treaty system covering climate change, are atmospheric carbon dioxide concentrations projected to continue to rise to perhaps as much as double or triple today's concentration, according to some models (IPCC 2001)? Why, in other words, are there effective regimes to address some global environmental problems but either very weak arrangements or none at all for others?

Of all environmental issues of international importance, ozone depletion, deforestation, and climate change have probably received more public attention across the globe than any others. These cases are all similar in several ways. They each became the subject of UN negotiations within the space of less than ten years, and all were influenced by the concept of "sustainable development", popularized by *Our Common Future* (WCED 1987). The negotiations themselves were based on the framework convention-protocol approach, a new and preferred technique of international environmental lawmaking. Indeed, the set of ozone treaties was the first set of truly global treaties that followed this approach. Unlike piecemeal treaties, such as various wildlife protection treaties, or a one-time self-contained global codification of law in a particular issue area, as in the Law of the Sea (Sebenius 1993), the framework convention-protocol approach allows parties to begin by defining the normative scope of a formal instrument in very general language with the intention of reaching greater specificity over time in a dynamic sequence of subsequent protocols (Sand 1992). The framework convention proclaims basic principles on which consent can be achieved; the envisaged protocols establish the method of achieving the general objectives through delineating more specific obligations.

Yet efforts by the international community to put effective arrangements in place to ameliorate these environmental problems range from very effective to almost completely ineffective. This book delves into all three of these issues, comparing the success story of the ozone regime with the failure to achieve binding global commitments on forests and the lack of effectiveness of the global climate regime. All three cases hold lessons for how to achieve realistic yet effective environmental solutions, given the state-centric, asymmetrical power structure of today's international system.

The Tragedy of the Commons as the Problem?

We can start with the question of which issue has more in common with the issue of climate change: ozone depletion, or deforestation? At first glance, the answer appears obvious. Stratospheric ozone depletion and global climate change are atmospheric problems. They are threatened by overuse in terms of human-produced gas emissions that affect the environmental processes that take place in the atmosphere and that are necessary for the survival of earthly life forms, including humans. For stratospheric ozone, the first and greatest perceived threat came from chlorofluorocarbons, or CFCs, although other human-produced gases made from chlorine and bromine compounds have also been found to interact detrimentally with the delicate layer of ozone in the stratosphere as well. The earthly climate, like the ozone layer, can also be affected by human-produced gases. Six gases are denoted in the Kyoto Protocol to the Framework Convention on Climate Change, and are thus given the collective label of "greenhouse gases" (GHGs),[2] but there are other GHGs as well, including CFCs, the first gas to be scheduled for elimination under the Montreal Protocol regime to protect the earth's ozone layer.[3] However, the effects of GHG emissions on the world climate are still uncertain and are likely not to be limited to "global warming" as was first posited.

Stratospheric ozone depletion and climate change are usually characterized as "tragedies of the global commons". Atmospheric resources belong to that category of environmental resources that come under the sovereign jurisdiction of no country; rather, everyone on earth theoretically has equal access to them. This similarity is significant for environmental goods. The fact that these are both "commons," or common pool resources (CPRs), means they are nonexcludable but divisible. In other words, it is difficult or impossible to exclude anyone from using the earth's atmosphere, but one appropriator's use of the resource diminishes the amount of the resource left for others. This becomes an issue if the atmosphere is used as a sink for wastes such as ozone-depleting chemicals and greenhouse gas emissions: as a result of human activity the amount of discharge of these gases into the atmosphere reaches a level that can no longer be absorbed by the atmosphere in its continual process of self-cleaning. This causes the atmosphere to be depleted as it becomes less and less able to provide the environmental services upon which humans depend. It is this juxtaposition of nonexcludability and divisibility that makes for the tragedy described by Hardin. Unless countries can come to an agreement that is not only precise but also binding and somehow enforceable on all parties—in other words, unless an effective international "regime" can be created to protect the atmospheric commons from being diminished by limiting access to it—the atmosphere is in danger of overuse, even to the point of irreversible loss.

In effect, this is what the Montreal Protocol is already accomplishing: limiting and, ideally, eventually halting the overuse of the ozone layer as a sink for emissions of CFCs and other ozone-depleting substances (ODSs). Because of the physical similarities between the phenomena of ozone depletion and climate change, and their common characterization as tragedies of the commons, some statesmen and academic analysts expected the success story of global cooperation for recovery of the stratospheric ozone layer to become a model for cooperation to halt global climate change. At least, the notion that negotiators endeavoring to address climate change should learn the lessons taught by cooperation on ozone depletion was promoted by numerous authors (Benedick 1991, 1998; Morrisette 1991; Sebenius 1991a; Tolba 1997). This was attempted, and today the 1985 Vienna Convention for the Protection of the Ozone Layer and its subsequent 1987 Montreal Protocol on Substances that Deplete the Ozone Layer are paralleled by the 1992 Framework Convention on Climate Change and its 1997 Kyoto Protocol, completed almost exactly 10 years after the adoption of the Montreal Protocol.

Despite the similarities between the cases of ozone depletion and climate change, they differ dramatically in the level of effectiveness of human efforts to address them. The quest to address ozone depletion has been remarkably effective in bringing about environmental improvement. It is predicted that the commitments made under the Montreal Protocol to halt emissions of CFCs and other ozone-destroying chemicals will result in a full recovery of the ozone layer within 50 years.[4] On the other hand, even though evidence that appears to confirm predictions regarding the effects of emissions of greenhouse gases is mounting,[5] efforts to tackle climate change through international policy-making have so far foundered. Indeed, in a physical sense, as emissions levels have generally continued to climb, the world has collectively moved further away from the goal of halting climate change since 1990—the year international policy-makers first began to address the issue. This raises the question of how comparable these two environmental issues really are. Perhaps the recognized similarities between depletion of the ozone layer and climate change do not capture the most important elements of these two environmental issues. This is important to recognize if our reading of history is to teach us the lessons that will bring about the outcomes that we want in the future. There are several differences between the cases of ozone depletion and climate change. Negotiations on the ozone layer attracted little interest at first, whereas from the beginning those on climate had more than 100 state participants, including developing countries that had been conspicuously absent in the early days of the ozone negotiations, as well as representatives of numerous international organizations and nongovernmental organizations (NGOs) (Breitmeier 1997, 104. Another difference lies in the activities associated with them. Although both problems are associated with industrialization, the production of ODSs such as CFCs was carried out almost exclusively by developed countries, whereas the developing world has more responsibility for activities that contribute to climate change, including agriculture and deforestation, and this will be even more true in the future (DeSombre 2002, 95–96).

However, the fact that both are commons has been perceived as outweighing all these differences. Conventional wisdom has it that it is the "commonness" of environmental commons, or CPRs, that makes problems of global commons difficult to solve. Because natural resources that fall within states' sovereign territories are more excludable, it is assumed that problems associated with their protection are manageable, whereas the overuse of global commons resources is assumed to be very difficult to address because they fall under no one state's authority. Two solutions to the dilemma posed by CPRs have traditionally been advocated: privatization—the development of private rights to the CPR, or the imposition and enforcement of limits to access by an external authority—or, in the words of Hobbes, a "Leviathan" (Hardin 1968; Ostrom 1990). At the global level, however, neither is there any supranational authority to manage the commons, nor is it easy to privatize global commons areas. Rather, addressing the global commons requires cooperation between sovereign states.

This has been achieved in the case of ozone depletion. Despite the fact that it is a global commons, the depletion of the stratospheric ozone layer has been addressed effectively through a series of ever tighter binding commitments that have been almost universally agreed upon by the sovereign states of the international system (UNEP 2005). On the other hand, the lack of success in attaining meaningful global cooperation on controlling the greenhouse gas emissions that cause climate change still seems to support the idea that overuse of the global atmosphere is in fact difficult to control. Indeed, the lack of effective results in the attempt to create a global climate regime has been called a "tragedy of the atmosphere" (Soroos 2001, 2).

A Different Kind of Tragedy

International processes to address the problems of climate change and ozone depletion lie at almost opposite ends of a conceptual spectrum of effectiveness. The crux of the problem must therefore be something besides the fact that they are both global atmospheric problems and both CPRs. In order to understand the reasons for the difficulty in achieving effective solutions to halt or at least mitigate climate change, it may be more useful to compare this case with another case of ineffective international environmental politics, the case of deforestation. At first, this issue might seem much too dissimilar to provide helpful insights into the climate change issue. Forests are not global commons; rather, they are found within national borders and come under the sovereign jurisdiction of one state. Therefore, any state where forestland is located is the ultimate arbiter of its fate, through its particular system of property rights. Forests are also natural resources, thus at least one of their values is monetized. Unlike atmospheric sinks, therefore, deforestation occurs due to both direct causes—in order to capture the value of forest timber—and indirect causes—as forests are cleared for agriculture or other development.

Thus, on the face of it, it seems that deforestation should be easier to address: forests have values that can be monetized, which should provide an incentive for their capture, and forests can be "privatized" by the states in which they are found so as to limit access to them. It is odd, then, that deforestation has taken place to such an extent that concern about the world's forests has risen to the international level. Odder still is that once this concern manifested itself in calls for international solutions, the international community completely failed to address tropical deforestation effectively. Although emissions of ozone-depleting substances have been largely halted, deforestation rates have decreased very little.

It has been posited by more than one participant in the ongoing process to formulate forest policy at the global level that perhaps resources under national jurisdiction cannot be addressed legitimately at the global level in the way a truly "global" issue such as ozone can, because they are not a shared resource.[6] This argument effectively turns the "tragedy of the commons" argument on its head, raising the possibility that national authorities with jurisdiction over a natural resource such as forests may not have an adequate interest in protecting them. This conflicts with the view of those who advocate establishing private property rights over natural resources for the sake of their protection. Yet the fact that at least some policy-makers perceive it as more difficult to achieve effective cooperation on resources under national jurisdiction than on global resources reflects a perception that it is sometimes difficult to convince countries with sovereignty over the resources in question to protect them or to halt their destruction. It is now well established that the interest of private owners in protecting a resource under their control depends on their individual discount rates, defined as the reduction in the present value of a good to an individual if the good will be obtained only at some point in the future. This will depend to a large extent on their particular circumstances. As an extreme example, the survival of a forest into the long term may be of little value to someone whose immediate survival is at stake; for this reason, the forest, which would otherwise be a renewable natural resource, may be "mined" (harvested at an unsustainable rate) in order to meet basic survival needs, directly (such as for fuelwood), by marketing the resource to produce an income, or through clearing the land for agriculture.

We thus have a paradox, in that the reasons offered for the lack of success in halting climate change and deforestation are exactly opposing. The fact that there are no property rights in the atmosphere is cited as the reason why the former problem is difficult to address, yet the fact that forestlands belong to states has failed to bring success in halting the latter problem.

A Different Answer

Perhaps the solution to this paradox lies beyond the question of whether climate is an atmospheric commons or forests are sovereign resources. I offer a framework that resolves the seeming contradiction presented by the international failure on climate change and lack of international success in combating deforestation and shows why climate change has more in common with deforestation than with ozone depletion. The only framework that can accommodate both the difficulty in obtaining cooperation on a global commons problem, such as climate change, and the difficulty in achieving protection of resources under sovereign jurisdiction, such as forests, is one based on an analysis of relative perceived costs and benefits of an effective solution. It is not the specific attributes of a CPR that make cooperation difficult as much as it is the effect these attributes may have on the costs of cooperation, in comparison with its potential benefits. The likelihood of obtaining effective cooperation on a global commons issue may or may not compare unfavorably with the likelihood of achieving effective action on more excludable resources. An effective outcome depends on whether all the costs involved, including any costs arising from the CPR characteristics of the global resource, are outweighed by the possible benefits of cooperation.

Starting with a cost-benefit framework carries an additional advantage over attempting to explain environmental politics beginning with a CPR model: It does not assume any common interest in provision of the environmental good in question. Common interest is the assumption behind the CPR concept, as well as the Prisoner's Dilemma (PD) game model, which is also commonly used to portray political difficulties in addressing environmental issues (e.g., Barrett 2003), but it depends on a positive cost-benefit assessment for all actors involved in negotiations for a cooperative solution to a particular environmental issue. Both the PD model and the tragedy of the commons metaphor describe a bargaining situation in which a cooperative outcome is clearly more desirable than a noncooperative outcome; yet the incentives for all players to act in their own strategic self-interest, or free-ride, tend to push them toward noncooperative behavior, thus producing a less desirable outcome. Therefore the key issue to be resolved is assumed to be how to enforce a cooperative agreement.

The PD model, shown in figure 1.1, carries the assumption that two actors have a common understanding that the cooperative solution (the C,C cell in the diagram below) is better than the status quo (the D,D cell), and that therefore the only political issue to solve is that represented by the D,C cell: the fact that each would prefer to let the other actor bear the costs of taking action.

Figure 1.1 illustrates the classic scenario of PD: Two prisoners have been caught for a crime but there is not enough evidence to convict them of more than a misdemeanor charge. The arresting officer suspects them of committing a felony but the only way to prove it is to get one or both to confess. This he does by separating them and offering each a plea-bargaining deal that would let off each if he rats out on the other, while his partner will receive the maximum sentence possible for the felony charge. This presents both prisoners with an incentive to confess, or "defect," which will bring the

		State B	
		Cooperate	Defect
State A	Cooperate	3,3	1,4
	Defect	4,1	2,2

Figure 1.1 Prisoner's Dilemma

most preferred outcome, four, if the other prisoner remains silent. This is combined with a disincentive to cooperate with each other by remaining silent because that risks the "sucker's" payoff of one, the lowest score possible. The result will be a mutual defection, which gives each a less preferred outcome (2,2) than they could have obtained had they both cooperated (3,3).

If a cost-benefit analysis is done regarding the choices facing each prisoner at the time he is asked to confess, the benefit to be gained individually from cooperating will be outweighed by the potential score from defecting no matter what the other prisoner does. Yet each prisoner in PD also knows that if he could trust the other they could get a collectively better score—and in any case an individual score that outweighs the score each will obtain from mutual defection—by cooperating. In other words, it is not a negative cost-benefit assessment that prevents cooperation in this scenario, it is simply lack of trust, probably justified.

However, for some actors, the cost of providing a collective environmental good can outweigh the benefits. In this case the question of enforcement of an effective agreement will not arise anyway. The agreement will be ineffective unless it contains provisions for enforcement or unless it is "self-enforcing"—in other words, unless it shifts the costs and benefits of agreement so that adhering to the agreement is in every actor's interest (Barrett 2003, 62; Keohane 2005, xiv). Furthermore, if agreement must be achieved by consensus, as is the case for the binding international treaties under discussion here, the question of enforcing an agreement that is not already "self-enforcing" in this way will not arise because not all actors will find it in their interest to include enforcement provisions. However, neither of these conditions requires an a priori assumption of common interest.

Taking the costs and benefits of effective agreement as our starting point brings the theoretical exercise more in line with the reality that the calculation of costs and benefits does not have to be similar for all actors in the system. Indeed, where the costs of agreement are seen to outweigh the benefits for all actors, the result is a "deadlock," as illustrated in figure 1.2.

		State B	
		Cooperate	Defect
State A	Cooperate	2,2	1,4
	Defect	4,1	3,3

Figure 1.2 Deadlock

In a deadlock situation there is no incentive to cooperate because the benefit from collective defection, or no cooperation, is perceived by all players to be higher (3,3) than the benefit of cooperation (2,2).

Even in a case where one actor believes cooperation will bring more benefit than defecting, such may not be the case for the other actor. In this situation an asymmetric deadlock is produced, as illustrated in figure 1.3.

Deadlock or asymmetric deadlock in global environmental negotiations may occur much more often than is actually acknowledged. Environmental issues, particularly at the global level, are generally characterized more by divergent interests than by common interests. Even though there usually appears to be a common interest in addressing the problem at hand, actors' rhetoric about the problem's severity and the need to act upon it is frequently not matched by commitments for action. The parties may not only disagree about which cooperative agreement is best but whether any agreement at all is desirable.

The act of coming to the negotiating table is not an indicator of common interest. In reality, there may actually be little interest in a cooperative outcome on the part of some actors, other than a desire to be seen talking about the problem. In other cases both parties may have an interest in the act of reaching an agreement (Lax and Sebenius 1986), but the minimum acceptable value of agreement, or "reservation price," for each party may not leave room for mutually acceptable agreements. In such a case the actors may only slowly realize that they are in a deadlock (Raiffa 1982).

Similarly, coming to the table may mean that at least one side hopes to "claim value" to itself at the other party's expense (Sebenius 1992a, 335) by linking the environmental collective good under discussion to an exchange for something more valuable to itself, such as debt relief, technology transfer, or another form of aid. The lack of an agreement on control measures for CFCs in Vienna in 1985 was the result of "value-claiming deadlock," because both the European Community (EC) and the United States insisted on controls that would favor their own differing domestic situations (see chapter 3).

		State B	
		Cooperate	Defect
State A	Cooperate	2,3	1,4
	Defect	4,1	3,2

Figure 1.3 Asymmetric Deadlock

Alternatively, some countries may participate in negotiations in order to prevent consensus on a deep level of cooperation from being reached among more like-minded countries, as Malaysia was accused of doing during global forest negotiations in the early 1990s.

Global environmental collective goods, particularly goods that will only manifest themselves in the future, share characteristics of intangibility, uncertainty, and unevenness of distribution, and they carry different meanings for different people; characteristics that adversely affect the helpfulness of an assumption of a common interest in their provision. Environmental goods are frequently intangible, in that benefits from their provision consist only of avoidance of a harm: fixing or ameliorating an already-existing problem, or restoring an environmental service to a former healthier state. To value such a benefit requires belief in a counterfactual argument as to what would happen in the absence of cooperation. Moreover, harm that is not felt in the present but only expected in the future is subject to the uncertainties of scientific predictive capabilities. Even ozone science, which ultimately resulted in a comparatively effective set of regulations to fight ozone depletion, gave policy-makers mixed messages the 1980s. The more distant is the future in which such benefits will be felt, the more subject they are to individual discounting, making it even more difficult to balance present costs with a promise of future benefits.

Furthermore, there is no inherent reason that the prospective benefits or costs of action will necessarily be distributed evenly, even if acute differences in opinion about what is equitable did not exist at the global level as they do (Hempel 1996). The fact that a problem is regarded as "global" means by definition that the producers of the problem are not the (only) victims; it is therefore possible that not enough benefits will accrue to the ones who must pay the costs of amelioration to induce them to take effective action unless some kind of positive redistributive measures are taken. In fact, producers of a problem who do not feel its impacts—such as upstream polluters of a river—will see no benefit, only a cost, to themselves for any effort to clean the river that does not compensate them for the costs they will incur. This question of different interests is related to the issue of the salience of an issue for different actors (Kahneman 2002). An environmental problem has more salience for, or commands more attention from, individuals who are more affected by its impacts, perhaps due to physical proximity. In this, salience is related to vulnerability, although an environmental problem may be salient even to some who do not experience direct vulnerability.

Rowlands (2001, 56) applies these arguments specifically to climate change: if climate change were really a tragedy of the commons, everyone would recognize an interest in cooperating, although incentives to defect would still exist. However, there are spatial differences in the impacts of climate change and in the costs and benefits of abatement of GHG emissions. Indeed, some states may well see themselves as possible winners if warming occurs, given that they might gain longer and more productive growing conditions or a more hospitable climate (Skolnikoff 1990). For these reasons some players may well prefer the status quo over an effective international agreement limiting emissions. That the benefits of emissions controls do not necessarily outweigh their costs seems clearly to be true for countries such as the United States, which "withdrew" its signature on the Kyoto Protocol under President George W. Bush. There is also good reason to think that oil-exporting states would rather have no convention at all than to defect themselves and "leave environmental protection to others" (DeSombre 2002, 1), as their economic interests are likely to be affected by other countries' reductions in the use of petroleum-based fuels in order to lower GHG emissions.

Finally, on a deeper cognitive level, postmodern scholars point out that the assumption that all parties concerned with a specific environmental problem share the same understanding of the problem is a "consistent mishap" in international environmental affairs. Wapner points to the fact that one person's "wilderness" is another person's home; what is valued by some in life as an endangered species may be valued more by others in death, as dinner or potential income (Wapner 2002, 169). Without common understandings, the question of common or overlapping interests should be explored, not assumed.

The question of common interest also depends on the goal of negotiations. Even parties to negotiations that otherwise favor an agreement may not favor an *effective* agreement. For example, some parties may want a global environmental agreement simply to appear to be doing something about the problem in question without actually being burdened by specific obligations (Humphreys 2005, 6; also see Parson 2003, 52). In this case, there would still be a deadlock with regard to an effective agreement.

Deadlock situations have generally been dismissed as uninteresting by students of cooperation (e.g., Snidal 1985b; Krasner 1991; Fearon 1998). According to Bernauer, the greater the difference in the players' interests, the less likely an agreement and the less likely that any agreement formed will be effective. Yet in fact, all three of the cases under study here began with an asymmetric deadlock, with some actors favoring the status quo over an agreement that they perceived would result in a net loss rather than a net benefit to them. Why did only one of these cases result in success? The key difference between them is who stood to benefit and who stood to lose from an effective agreement, and how these interests mattered for the outcome of global environmental negotiations.

Leaving aside any assumption of common interest, I offer a two-tiered explanation of differences in the outcomes of negotiations to create a global environmental regime. First, leadership is needed. In an asymmetric international system, where variation in countries' economic power is extreme, and under conditions of divergent preferences, in which some states perceive that an effective agreement would be more costly than beneficial to them, an effective agreement can be reached if the leader—the state with the greatest capability—manipulates the preferences of anti-agreement states. This can be done either by raising the value of agreement for them through promises of rewards or by lowering the value of no agreement through threats of sanctions. Given the United States' current position as the most powerful state of a very asymmetric international system, an effective outcome depends on US leadership. This leads to the second point: the lead state's willingness to lead depends on its expected costs and benefits from an effective agreement, including the costs of manipulation of other states' preferences.

In the following pages, I present a typology of the range of potential costs and benefits that the United States may expect to incur with the negotiation of an effective global environmental agreement. By using this typology of the United States' interests in the areas of ozone depletion, global deforestation, and climate change, it is possible to gain a deeper understanding of the similarities and differences in the United States' position in these cases and, ultimately, of the reasons behind the different outcomes of efforts to address these global issues. Thus, in a case where the US stood to gain a net benefit from the formulation of an effective agreement, the outcome has indeed been a highly effective agreement. Where, however, the US stood to be a net loser given its cost-benefit equation, agreement either has not resulted or has been ineffective.

Chapter 2 considers the current state of theory on environmental regime creation and explains my approach in detail, showing how it expands our understanding of global environmental cooperation. Chapters 3 and 4 will consider the ozone success story in depth, first in terms of the original efforts to control CFCs and then in terms of the contentious negotiations that have taken place on some other ozone-depleting chemicals and that have generally received much less scholarly attention than those pertaining to CFCs.

Chapter 5 will then switch gears, to consider the failed case of negotiations for a global forest convention, efforts for which began in 1990 and continue to this day. We will then take up the question of climate change in chapter 6. Although global negotiations on climate change succeeded in producing a convention and a protocol, thus emulating the convention-protocol approach first used in the ozone negotiations, the lack of effectiveness of the global climate regime more closely resembles the failure of the forest negotiations. The climate change case represents an interesting test of the explanatory power of my approach in that it covers an issue in which the United States has never taken a lead in advocating effective global environmental action and where the United States has been a vocal opponent of many international initiatives to address it, including the Kyoto Protocol. However, its greater importance lies in the large looming threat that unmitigated climate change poses to humanity and to our planet, as consistently predicted by overwhelming scientific consensus. An examination of the United States' cost-benefit structure pertaining to effective environmental cooperation, as offered by my framework, identifies potentials for shifts and points of leverage that may well be key to addressing this looming threat effectively.

In the final chapter, chapter 7, I consider the cases together to draw conclusions on the best way forward. The pressing issue is how to improve cooperative efforts in the future, in order not only to raise their chances of success but also to ensure that the limited resources available to bring all needed actors together to develop common interests in effective agreement are expended as efficiently as possible.

CHAPTER 2

A Different Approach to Understanding Environmental Regime Creation

"The environment" became an issue when the effects of the deterioration of many environmental services and its risks to human health began to be perceived. This first took place mainly in developed countries, the obvious connection being that it was there that human activity had been undertaken on such a large scale, thanks to industrialization, that environmental services could not cope. Thus, air and water pollution and land contamination problems began to be addressed at the domestic level in numerous countries.

In the 1970s, however, new environmental issues arose that no one country could address successfully on its own. They rose to the international agenda as policy-makers realized that efforts to control or halt the effects of these larger scale problems necessitated cooperation between sovereign countries. By the 1970s, a number of environmental treaties had been negotiated, mainly between developed and other countries that found a common interest in protecting the resource concerned. Then ozone depletion necessitated a new kind of global treaty between countries of widely disparate interests. The Montreal Protocol was produced, and policy-makers and environmental scientists alike took to heart the apparent lessons it offered for policy-making on other environmental problems at the international level. In the space of five years after the adoption of the Montreal Protocol at least five other global environmental conventions were proposed, including not only those on climate change and deforestation that are examined here but also conventions to control transboundary movement of hazardous wastes, to halt the loss of biological diversity, and to combat desertification.

This was a (brief) golden era of global environmental cooperation, and it became a popular focus for academic investigation by numerous scholars of international relations. As time has passed and later attempts to cooperate on addressing other international environmental problems have generally been less successful, enthusiasm for the project has waned, and along with it so has academic enthusiasm for studying the subject of "environmental regime creation."

But why should scholarly study of efforts to achieve international environmental cooperation be less popular now? There is no logical reason why this should follow. Lack of cooperation on environmental issues indicates an area in which scholarly contributions toward increased understanding of the difficulties of cooperation in this area would perhaps be welcome. Perhaps it is because existing theories have been unable to explain the general failure of the international community to grapple successfully with the enormous environmental problems that confront us. The fact that environmental policy-making differs substantively from policy-making in other areas, such as security or economics, means that existing theories of international cooperation—particularly those associated with "collective action" as it has been conceived—do not necessarily accommodate the circumstances of environmental policy-making.

First, environmental problems differ from issues of security or economic cooperation in that they are ongoing negative trends—if no cooperative action is taken, the status quo will progressively deteriorate—even though effects may only be felt in the medium to long term. This is sometimes, though not always, true of economic or security problems but it is an inherent characteristic of environmental problems.

Second, the fact that cooperation on environmental issues can only produce the intangible benefits of ameliorating a harm in the present or avoiding an uncertain harm in the future makes environmental problems unable to compete in importance with security or economic issues that are much more immediate in their effects and for which cooperation may bring more tangible benefits, such as the promise of positive improvement in economic position that economic cooperation brings. It also increases the likelihood of widely divergent interests among the states involved, as noted in chapter 1. Moreover, the more distant in the future the impacts of environmental deterioration are predicted to be, the less likely it is that such predictions will capture attention when competing with more immediate interests. Environmental deterioration whose effects may only be guessed at and only threatens harm in the much longer term does not pose an immediate threat, thus differing from security issues; in addition, it will be difficult to perceive any benefit in the near-term from halting the deterioration, but any interest in addressing it must still compete with states' interest in pursuing short-term economic welfare that can be predicted fairly accurately and measured in monetary terms. "Borrowed" theory cannot encompass these complexities.

These theoretical difficulties may be associated with a waning of scholarly interest in environmental collective action problems, apart from a growth in critical analyses of the environmental problematique as the inescapable outcome of a bankrupt international system (see, e.g., Litfin 1994; Vogler and Imber 1996; Elliott 1998; Paterson 2000; Stevis and Assetto 2001; Broadhead 2002; Lipschutz 2004). Yet, if there is to be any hope of ameliorating the situation of environmental deterioration facing humanity, we still need to understand failures and successes in cooperative efforts in the world as it currently exists, given the unlikelihood that wishing power hierarchies to disappear will have the desired effect. This chapter therefore briefly sketches the current state of academic discussion surrounding the creation of environmental regimes and outlines my approach to greater understanding. It is only through a deep understanding of the factors that truly explain outcomes that scholars may eventually be able to predict outcomes in the future; it is hoped that the framework presented here will not only offer that possibility but will also enable those who want to do so to find ways to boost the odds of effective regime creation.

The Question of Effectiveness

I start without an assumption of common interest.[1] The question is how preferences may change in a situation of asymmetric deadlock—in other words, a situation in which some states want to take effective action at the international level to ameliorate an environmental problem but others oppose such action.

Before analyzing in more depth why any environmental issue should be characterized as one of deadlock, let us first consider what we mean by "effective". Many international relations scholars have expended pages trying to assess the effectiveness of international regimes. This term is used in various ways in the literature on environmental regimes, frequently with little attempt to specify its meaning in different works. Yet effectiveness as a term is highly ambiguous. There are three issues surrounding "effectiveness" that need clarifying here: First, do we mean effectiveness in enforcing an agreement or effectiveness as written into the agreement when negotiated? This is key, because this question underlies the fundamental debate over how to model the negotiating scenario in the first place. Related to this is the question of whether effectiveness is measured in terms of behavioral changes or amelioration of the environmental problem at issue. For most studies of cooperation in other areas, effectiveness in influencing behavior, particularly the behavior of states, is the goal; for environmental issues this definition is not enough of an indicator that the environmental problem itself is being effectively addressed. Finally, is pushing for a binding international environmental agreement synonymous with desiring an effective agreement? This might seem obviously so, but, as noted in chapter 1, states bring many different interests to the negotiating table; even those states that are the initial proponents of an international agreement may not have solving an environmental problem uppermost in mind (Humphreys 2005).

Effectiveness as Enforcement

The early study of effectiveness can be likened to a study of enforcement, given the modeling of environmental cooperation as a PD (Prisoner's Dilemma) game or a tragedy of the commons. This is because these models focus on enforcement problems, under the implicit assumption, once again, that both players agree to the benefits to be had from mutual cooperation (the CC outcome) and that both prefer that to mutual noncooperation. The question that must be answered for a problem modeled as a Prisoners' Dilemma is how to keep players in the C,C cell of the matrix, with a score of "3", given that each has incentives to defect in an attempt to get a higher payoff of "4".

Translated into international relations, how do you get players to honor a treaty they have signed, given the incentive that all players have to try to "free-ride," attempting to reap the benefit of everyone else's willingness to adhere to their commitments while not living up to their own? Proposed solutions include encouraging compliance with commitments by creating institutions to increase transparency and provide enforcement mechanisms or by reducing the size of the group participating in the cooperative endeavor, or have focused on the usefulness of having a hegemonic power with a particularistic interest in supplying the good even if others free-ride (see, e.g., Hardin 1982; Haas, Keohane, and Levy 1993; Downs, Rocke, and Barsoom 1997, 1998; Keohane 1997; Botcheva and Martin 2001).

A large limitation in beginning with such a model is that it does not address the negotiation process itself (Krasner 1991; Fearon 1998). Game models such as PD cannot represent dynamic distributional bargaining problems. Studies of cooperation that begin with PD generally do not pursue an analysis of the negotiations themselves but, rather, focus on the common interests that the parties already have.[2]

In line with these critiques, many recent offerings in this vein have shifted from the question of regime creation to the question of regime implementation. This fits well with the conceptualization of environmental goods as CPRs (common pool resources) and PD problems, given these models' focus on enforcement and given the fact that regimes already established can be assumed to be based on some degree of common interest. Young (1997), Wettested (2001), and Miles et al. (2002) explicitly address effectiveness as institutional effectiveness in enforcing commitments.

There are at least two problems with defining effectiveness so narrowly, however. First, as Kütting (2000) points out, "effectiveness" in the sense of successful cooperation does not necessarily result in a successful solution of the environmental problem in question. In 1993, Haas, Keohane, and Levy argued that there is great difficulty in assessing institutional effectiveness in ameliorating environmental problems. However, Miles et al. attempt to do precisely that in their study, seeking to assess outcomes of environmental institutions not just in terms of changes in behavior but also in terms of their environmental impacts. Second, however, a focus on already established institutions denies authors working in this area an opportunity to compare the circumstances and processes that generated those cooperative arrangements with environmental situations that did not. To their credit, Miles et al. recognize this limitation and acknowledge that study of such cases might shed light on the causes of failure.

In real life, the institutional setting within which effectiveness in obtaining environmental results is measured is determined by the actors involved in its creation. The structure of an institution is fundamentally dependent on actor preferences—and in particular the distribution of power over the configuration of interests—during its negotiation. One may wonder whether other factors such as the distribution of power over preferences during initial negotiations to establish a regime may go further toward explaining its ultimate effectiveness than can simply measuring how well its obligations are implemented, with the "institutional setting" variable taken as a given.

None of these studies on effectiveness as enforcement asks whether some actors might in fact perceive more benefit, or less cost to themselves, from maintenance of the status quo, and all of them depend upon a prior recognition and acknowledgment of common interests. However, reaching a point at which all actors acknowledge a common interest and can agree to a common goal is not straightforward. This is all the more true for environmental goods, given their characteristics of intangibility, scientific uncertainty, and unevenness of distribution.

Effectiveness in a Negotiated Outcome

A separate stream of theorizing on international environmental cooperation has advanced understanding by seeking to identify what may account for the ways in which configurations of interests are formed, the extent to which interests are held in common, the reasons behind divergence or convergence of interests, how interests may shift, and how all this affects the outcomes of negotiations. This literature may be dated to Peter Haas's (1989, 1992a, 1992c) theory on epistemic communities—scientific networks that produce common scientific knowledge and use it to convince policy-makers in opposition to other interest groups—and to Sprinz and Vaatoranta's (1994) oft-cited article positing "the interest-based explanation for international environmental policy."

Epistemic communities are defined as transnational groups that are bound together by a shared set of causal and principled beliefs theory, including shared notions of validity; in other words, scientists (Zürn 1998). Rather than assume that all potential parties to cooperative action have "generally consonant interests" (Hardin 1982, 84), Haas acknowledges that the United States exercised "hegemonic leadership" in the ozone case, arguing that this leadership role was a result largely of scientific consensus and the extent of the epistemic scientific community's influence on the US government (see Haas 1989, 1992a, 1992c). The claims that epistemic communities theory makes for the primacy of science and the scientific community have been the focus of much attention but also sharp criticism, particularly about the theory's lack of clarity as to the ways in which the knowledge community and those in power influence each other. Moreover, numerous studies have raised questions about the ability of the scientific community to influence all negotiations in the same way, given the existence of "adversarial science" (Susskind 1994, 65) and the fact that increasing scientific certainty is not necessarily associated with greater interest in cooperative action unless new information or new tests lead to predictions at least as threatening as those made earlier (Downie 1996).

Sprinz and Vaahtoranta (1994; see also Sprinz and Weiß 2001) also address variation in interests across countries, attempting to explain cross-national variance in support for international environmental regulation using two interest-based factors. Their factors of interest are a state's "ecological vulnerability", or the environmental threat it faces, according to the relevant scientific community, and its "economic capacity", which they define as the economic costs of abatement for individual countries. They provide a matrix showing how differing levels of these two factors interact. Although their framework brought a major leap forward in understanding in 1994, there are several deficiencies. Sprinz and Vaahtoranta do not attempt to compare their "abatement costs" with the costs of inaction, nor do they capture the fact that some countries may benefit economically—relatively or absolutely—from abatement, by, for example, winning a competitive advantage in abatement technology. Equally importantly, their use of the terms "costs of abatement" and "economic capacity" as if they are interchangeable ignores the fact that some countries will be able to meet a certain level of abatement costs while others will not, depending on the economic resources available to them. How should abatement costs be compared across countries of different economic capacities? Sprinz and Vaahtoranta recognize that feeling the effects of a problem is a prerequisite to an interest in addressing it; however, their dual-factor explanation implies that countries that have a problem but few resources do not have an interest in "pushing" for international regulation to ameliorate it.

Finally, although Sprinz and Vaahtoranta purport to explain "international environmental policy", their study does not actually pursue whether an agreement is likely to be reached if a set of parties to a negotiation have the differing interests they identify. This might depend on other factors such as the strength of the states pushing for regulation. Although motivation is a necessary condition for international environmental regulation, it is not sufficient unless those countries with such a motivation can lead others into cooperation, given a group of countries with conflicting interests and different capabilities. This may lead into questions of different states' relative capacity to influence others, but such questions are not addressed by Sprinz and Vaahtoranta.[3] Nevertheless, their study provides an important first step toward a fuller understanding not only of the conditions under which science may ultimately lead a country to support international cooperation to ameliorate environmental degradation, but also of related domestic variables that might be affected by the process of bargaining that was neglected in the earlier collective action studies.

Two recent offerings by DeSombre (2000) and Barrett (2003) significantly advance the evolution of theorizing on the implications of differing interests for the development of international environmental regimes and in effect form a foundation for the present work. DeSombre examines international environmental cooperation as the result of "internationalization" of domestic policy. DeSombre's study targets the questions of which domestic policies are chosen for internationalization and, then, the extent to which the state wanting to internationalize its policy succeeds in convincing other states ("target states") to adopt similar regulatory policies. In order for a state to be motivated to support or push for international environmental policy there must be both strong environmental as well as economic incentives for domestic actors. If there is no economic benefit possible for strong industrial actors, the chances of achieving effective environmental regulation are small to nil. For DeSombre, though, the economic interests involved are broader than Sprinz and Vaahtoranta's costs of abatement. If industry is already affected by the costs of meeting domestic regulations it will have an incentive to push internationalization of those policies in order to make targeted states bear the same costs as they do for environmental protection; alternatively it will push for international regulations to keep out goods produced by target states not subject to those costs (DeSombre 2000, 169).

Unfortunately, DeSombre (2000, 40) makes no attempt to develop a more general framework for explaining when her coalition of environmentalists and industry ("Baptists and bootleggers") is likely to form, nor to consider whether outcomes might differ if another state tried to internationalize its own domestic legislation against the opposition of the United States and whether those outcomes might vary in effectiveness. Moreover, her analysis of industry's interest intertwines its desire to "level the playing field" of environmental regulation internationally with the possibility of positive particularistic benefits that industry may be able to obtain through internationalization of domestic policy, without treating them as independent effects. This misses a key point of Oye and Maxwell's (1994) findings, with regard to the ozone case in particular, that industry pursued not just a leveling of the playing field but particularistic rents and subsidies. The difference in success and failure of environmental regulation, according to these authors, depends not on whether it is domestic regulations being pushed forward but on whether tangible benefits are conferred upon those that are regulated.

The facts DeSombre presents from the ozone case, used as illustration, are a little suspect in any case, given the fact that international regulation of ozone-depleting substances achieved under the Montreal Protocol does not look like the US regulations that the United States attempted to internationalize, and the fact that the CFC industry objected to international regulation for a full nine years after the original US regulations came into being. Indeed, it was only after the United States changed its position on international CFC regulation to call for an agreement that did not actually look like its domestic regulations that US industry eventually dropped its opposition (see chapter 3).

Barrett (2003) starts with the international environmental negotiating context itself, offering a way of strategizing in order to achieve the most effective treaty design. Barrett points to the need to restructure incentives to enforce participation as well as compliance. Again, however, he starts with the traditional PD framework, which forces a strained interpretation of the negotiating that goes on as lying within the scope of enforcement issues. It also assumes common interests, thus hindering Barrett's ability to see the larger issues at stake: "Recognizing that it is in their joint interests to [abate pollution], we might suppose that the two countries will negotiate an agreement which alters the payoffs in such a way that each state's own interests compel it to play 'Abate' (in other words, to cooperate)" (2003, 62). Barrett conceptualizes the main shortcoming of the Kyoto Protocol as one of enforcement, because negotiators did not deal with this issue in a timely enough way because of a mistaken view among negotiators that it could be added on later. He accepts at face value what participants in Kyoto told him, that the issue was put off because "you cannot solve every aspect of this problem in one stroke" (2003, 360–362). The question of whether the enforceability of the commitments made might run counter to the interests of powerful actors does not seem to have occurred to him.

Granted that an agreement has to be self-enforcing in order to function. This requires that all parties benefit more from an agreement than from the status quo, a point already made by Keohane in 1984 in his groundbreaking work, *After Hegemony* (which also assumes a common interest in an agreement that can be effectively enforced). Neither Barrett nor Keohane come fully to grips, however, with the fact that the effort to design a treaty incorporating such incentives is dependent on the bargaining process, or the fact that bargaining depends on leverage, which has much to do with the power structure involved. The achievement of effective environmental treaties depends on the pro-agreement coalition having not only the desire for an effective treaty in a given issue area but also the power, or capability, to pursue the strategies Barrett lays out. This is all the more true, as I outline in the chapters to come, for the more usual situation in which there is no common interest in ameliorating the environmental problem at hand.

The Effectiveness of Agreement

In light of the variation in what "effectiveness" is taken to mean in the various studies that purport to analyze the effectiveness of international environmental cooperation, I propose my own definition of effectiveness in terms of the outcomes of bargaining processes. I leave aside questions on implementation because implementation is only as good as the quality of the commitments being implemented (Bernauer 1995). First, an ideal effective negotiated outcome must be a legally binding international agreement, or set of agreements, covering a particular environmental issue. An effective agreement is thus distinguished from nonbinding "soft" law. Although Guppy (1996) and others have argued that there is little difference between binding ("hard") international law in the form of weak framework conventions and "soft" law in the form of declarations or statements of principles, there are at least two critical differences: First, while legal conventions may be as vague in their obligations as soft hortatory declarations, only binding commitments can be followed by protocols that are intended to set out specific obligations. Second, only legally binding commitments, no matter how weak, can have mechanisms to enforce compliance (Davenport 2005). Significantly, it is binding treaties that are still the primary means for achieving international cooperation, not only on global commons and shared environmental resources but on any issue for which the need for cooperation at the international level is considered urgent. For any global environmental issue that is considered a crisis, one can assume that the appropriate mechanism for international cooperation to address the crisis would be the same mechanism used to address other urgent matters requiring international cooperation: a binding agreement, with as universal a participation rate as possible.

Of course, a binding treaty still may not be effective. Ultimately, effectiveness must mean a physical improvement in the environmental problem being addressed, as per Kütting's explication. Unfortunately, it is difficult to judge recent treaties fairly given that it may take decades for the environmental problem in question to be reversed. There are proxies, however, that can be used to judge treaty effectiveness even in the absence of knowledge about the extent to which implementation will provide the desired environmental good. These include the commitments made and the treaty's language regarding compliance and enforcement. In addition, the rate and quality of participation in the treaty is an important aspect of effectiveness, as is the treaty's relationship to other agreements. Individual aspects of effectiveness within each of these categories are listed in table 2.1.

Table 2.1 Criteria of Effectiveness

Commitments	*Compliance*	*Participation*	*Treaty relationship*
precise targets	specified monitoring or review procedures	universal participation	language specifying relationship to other instruments
timetables for their achievement	body established to review implementation	producers of the problem	superiority to other instruments
firm, concrete commitments rather than "indicative goals"	specified dispute resolution body	potential providers of financial resources to assist implementation	specified procedure for resolving conflicts with other instruments
	enforcement procedures	potential providers of technology to assist implementation	specified body to resolve conflicts with other instruments
	sanctions for non-compliance		
	mechanism for assisting implementation		

These elements refer to the following questions:

Commitments: Does the agreement specify precise behavioral targets to be met by the parties? Are there timetables specified for when these targets should be met? Are the targets intended to represent firm commitments or simply indicative goals? These factors must be considered cautiously, of course, as timetables may, for instance, appear to restrain countries from doing something they would not do anyway.

Compliance mechanisms: Does the agreement lay out procedures for monitoring and/or review of compliance? Is a body for review of implementation established or identified? Is a body established for dispute resolution? Are any sanctions for noncompliance specified in the agreement? Is there any mechanism for providing assistance, such as financial or technological aid, to countries in order to help them implement the agreement?

Participation: Do the parties to the agreement[4] include the major producers of the environmental problem being addressed? Do the parties include potential providers of financial resources to assist implementation or potential providers of technology to assist implementation? How universal is participation, given that these are all global issue areas?

Treaty relationship: Does the agreement contain language specifying the nature of its relationship to other agreements in overlapping or related issue areas, if any? If so, is the agreement intended to supersede other instruments or rank as "superior"? Is any procedure for resolving conflicts with other instruments specified in the body of the agreement? Is a body established or specified in the agreement for resolving conflicts with other instruments?

Does a weakness in any of these areas indicate that an agreement will not be effective in addressing an environmental problem? Yes, if the agreement lacks commitments that will really alter behaviors seen as contributing to the problem and if the agreement is itself necessary to address the problem. The lack of enforcement mechanisms need not, on the face of it, prevent environmental improvement from taking place, but given the human propensity to free-ride, enforcement is still an issue. But the final two categories also indicate areas of potential weakness in a treaty that would prevent it from being effective in ameliorating the environmental problem it is intended to address. If a treaty is negotiated that requires parties to alter their behavior significantly, at a cost to themselves, then it may be weak in having few countries sign up to it. Barrett (2003, 356) identifies this problem as a "trade-off between the depth and the breadth of cooperation"—the need to balance the threat of noncompliance with that of nonparticipation—and notes that in the negotiating process countries must sometimes choose between "broad but shallow" and "narrow but deep" treaties. Finally, if a treaty requires strong commitments of its parties but is then superseded by another treaty placing incompatible obligations on its parties, this too will weaken the prospects for effective environmental cooperation.

From Asymmetrical Deadlock to Effective Agreement[5]

Having now established a definition of effectiveness that can apply to all three of the cases we will delve into in later chapters, we now return to the question of what is needed in a situation of asymmetrical deadlock in order to achieve an effective agreement by this definition. Oye (1986, 6–7) has asserted that this can never happen: "If at least one actor prefers nominal mutual defection (DD) to nominal mutual cooperation (CC), 'policy coordination' cannot lead to mutual gain; the term 'cooperation' becomes inapplicable. Symmetric and asymmetric games of Deadlock fall into this category. . . . Where deadlocks exist, the term 'cooperation' is devoid of meaning and conflict is inevitable."

But this may or may not be true; a bargaining process might potentially shift preferences so as to make Oye's "inevitable" conflictual outcome more cooperative. Indeed, in a discussion of arms races in the same volume, Downs, Rocke, and Siverson (1986) point out that although "negotiation may seem to hold little promise for reducing the intensity of a Deadlock . . ., [it] offers the possibility of linking that game to other issues in such a way that the net marginal benefit that each gains from cooperating within this larger game is greater than that of defection." This possibility has received very little attention, however, even from these authors themselves.

I argue that deadlock can be overcome if the side preferring agreement is able to change the other side's perception of the value of the agreement outcome relative to the no-agreement outcome. A necessary condition for a party's agreement is the prospect that the agreement will have greater perceived net worth than that party's best course of action without agreement. Where this condition is not met, a party for whom an effective agreement has net positive value may be able to manipulate the preferences of other parties toward the agreement outcome, either through enhancing the potential of joint action or through lowering the value of the other party's alternative to agreement. Manipulation or coercion of the perceptions and preferences of one side by the other side is most likely to be associated with issue linkage—incentives, such as promises of rewards, or threats of punishment.

The feasibility of manipulating preferences or resisting such manipulation depends on the bargaining power of each side of a dispute. Bargaining power depends, to a large extent, on how easily one or both sides can "walk away" from the agreement. For instance, in an asymmetric deadlock over acid rain, an upstream state, which perceives much to lose but seemingly little to gain from agreement, has much more bargaining power than a downstream state that needs an agreement in order to halt the harm being felt there (see Mitchell 1999b, 3–4). Whether or not an actor can walk away depends on whether the other side has something the actor wants and how much the actor wants it. Multilateral negotiations are thought to differ somewhat from two-sided negotiations inasmuch as they add the possibility of fluctuating alliances and coalitions and third party mediators. However, in cases that are affected by North-South politics, as all of the cases considered here were, the coalitional boundaries between developed and developing countries were for the most part drawn in concrete. Because there were therefore few "congeries of dyads" in these negotiations, there is little need to worry about the complexities that might arise from them (see Zartman 1994, 3).

Bargaining power also depends on the extent of the resources available for use to bring about changes that cause preferred outcomes,[6] such as resources that permit a party to punish or reward another party for its behavior. In an asymmetric deadlock situation, manipulation of preferences for effective agreement depends on the relative bargaining power of the party favoring agreement, in terms of the resources that can be used to change perceptions, as opposed to the bargaining power of the other party to resist such pressure and either change the deal or walk away.

This raises the question of how power asymmetry affects preference manipulation. Zartman notes that there has been too little work on the direct implications for the parties on either side of a power asymmetry, although Lax and Sebenius themselves acknowledge that "greater resources may translate into greater capacity to impose sanctions" (1986, 253). The structure of the international system precedes the process of any international negotiations in determining outcomes. What may be true for a negotiation among roughly equivalent actors—at least in the sense of all having the power to walk away from agreement—cannot be assumed to hold true for actors with wide disparities of power resources. For example, the global situation differs from that of transboundary acid rain negotiations in that while the latter may take place between states of roughly equivalent size or power resources, negotiations in global issue areas by definition include countries representing the full range of sizes and power resources.

Given an asymmetric deadlock at the beginning of negotiations, effective agreement depends on the pro-agreement side's capacity to manipulate the anti-agreement side's perception of the relative values of the agreement and no-agreement outcomes. The greater the capability of the pro side, the more effective the agreement can be. The ultimate implication of this is that the pro coalition must include the most powerful state in order for effectiveness to be increased through manipulation. First, if the state with the most capability is in the pro coalition, this opens the possibility of using some of its resources to meet any costs of coercion. On the other hand, if the state with most capability is not in the pro coalition, not only are its resources not available for such use but manipulation of that state's preferences then also becomes an issue for the pro coalition; this poses an additional cost on top of the fact that the pro coalition already has fewer resources to use for such linkage.[7]

This may be exacerbated by differences in perceptions of absolute dollar amounts. What is perceived as inexpensive for some may be exorbitantly expensive for others, depending on states' attributes. Although sometimes differences in the physical characteristics or needs of states can lead to effective issue linkage based on a trade of goods valued differently by the different states (Sebenius 1983; Morgan 1994), differences between states in terms of their level of "economic vulnerability"—in other words, dependence on imports or on external markets for domestic products (DeSombre 2000, 161)—or in their levels of capabilities more generally may influence their perceptions of the absolute dollar amount of the costs and benefits of agreement and particularly of issue linkage, with the effect of favoring the initial interests of the less vulnerable state. For instance, Martin argues, with regard to sanctions specifically, that it does not matter if the absolute dollar amount lost by the United States in imposing sanctions on Guyana exceeds the cost to Guyana; what matters is the relative cost to Guyana of bearing the costs of sanctions versus the costs of changing its behavior to conform to the sanctioning state's wishes.[8]

This point can be extended to include consideration of the reverse situation: in the extreme hypothetical case in which Guyana might want to impose sanctions on the United States, it is obvious that Guyana would probably never be able to impose large enough sanctions to change the United States' perceptions of costs and benefits of conforming to Guyana's wishes. Any leverage that Guyana might at some point acquire over a larger state, as per Schelling's famous argument that "in bargaining, weakness is often strength" (1956, 282) would have to arise from the particular circumstances of a bargaining situation. Such leverage points have sometimes enabled small countries to exercise influence over the United States, as when Norway was able to leverage US acquiescence in stricter methyl chloroform controls in 1992, as discussed in chapter 4. However, this is more likely when both sides clearly prefer agreement over no-agreement alternatives, not in an asymmetric deadlock, a point that Schelling ignores. Furthermore, any strategic advantage that the weaker side might obtain in a given bargaining contest is further eroded by the fact that such tactics would not be available in an ongoing and systematic way, and if they were the more powerful members of the community could be expected eventually to find a way to destroy the technology that enabled such tactics to be used (Knight 1992, 131). Weak anti-agreement states may have enough defensive bargaining power to walk away and thus prevent agreement, despite their vulnerability to manipulation of their preferences. However, weak pro-agreement states are unlikely to have enough coercive bargaining power to manipulate other countries into favoring an effective agreement, because they would feel the costs of attempting to coerce cooperation more than the economically dominant country would feel the cost or benefit resulting from any attempted manipulation through sanctions or incentives. Indeed, the idea that the United States cannot be manipulated in this way has been confirmed by a US negotiator:

> Other countries don't try to use promises or threats to try to get the US to change position (we're immune to these anyway). More often people try to get us to change our position by insulting us, embarrassing us, or accusing us of God knows what underhand motives. Definitely this can make things awkward and harder for us by poisoning the atmosphere of the room in an anti-US way—but at the end of the day it's water off a duck's back to an experienced negotiator.[9]

It almost goes without saying, too, that putting political pressure not to block consensus does not work on the United States as it does, for instance, with Japan (see chapter 4):

> The US is not susceptible to that pressure but all our little allies are (Canada, New Zealand, Australia). No matter how strong they start, they fade without us. But we negotiate on principle. If we can't live with it, we don't. But, remember, a skilled negotiator knows exactly what he can or can't live with and how to manipulate text so he only rarely has to block consensus.[10]

For all these reasons, in a situation of great power asymmetry and asymmetric deadlock, any attempt by a pro coalition that does not include the economically dominant state to manipulate the perceptions of an anti coalition that does include the dominant state is likely not to be successful or to take place at all. But movement from an asymmetric deadlock to effective agreement through manipulation of perceptions of the zone of agreement requires that the actor with the asymmetrical share of power resources, or capacity, also have the will to coerce agreement. If those two traits do not intersect in the same actor, no effective regime will result.

This notion of leadership as necessitating both capacity and will corresponds to that of other authors. Most and Starr (1989, 31, 34), for instance, dichotomize choices in international relations into questions of opportunity and willingness. Opportunity, as they see it in the overall context of international relations, encompasses the distribution of capabilities, while willingness is closely related to a decision-maker's calculations of advantage and disadvantage—of cost and benefit—that decision-makers consider on both conscious and unconscious levels. Snidal states that leadership "depends on a conjunction of resources and initiative" (1990, 345). Morgan (1994) similarly breaks down bargaining power into capability and resolve in the context of international crisis negotiations. Finally, Underdal sees what he calls "coercive leadership"—the use of threats of punishment and promises of rewards—as a function of two major determinants: (1) capabilities/structural position and (2) behavior—"a certain minimum of effort" (1994, 182). These two factors correspond roughly to the power and will I specify here as necessary for preference manipulation, except I argue that the will to lead must precede the behavior; the primary question is whether a state will have the willingness to attempt leadership, not simply whether it will succeed given an assumption of willingness.

Thus, in the case of acid rain, for example, if the state with a desire for a cooperative outcome—the downstream state—is more powerful than the upstream state it will likely be able to manipulate the preferences of the upstream state in order to achieve an effective regime to stop the harm being felt. If it is less powerful than the upstream state, however, this will not be the case. In the case of global issues, as opposed to the transboundary problem of acid rain, the requirement for capacity and will becomes more acute as the valuations of many states come into play. In the multilateral context of the global system, this once again translates into a requirement for the most powerful state in the system to be a member of any pro-agreement coalition in order for effective agreement to be achieved.

The United States is generally recognized to be the most powerful state of the international system at the present time. Underdal, for instance, identifies the prime candidate for leadership as a state with a predominant position within the system of activities in question and with a high score on an overall power index. He notes that since World War II "the US has probably more often than any other state found itself in a position enabling it to exercise coercive leadership" (1994, 187). Snidal, in examining its role in international organizations covering other issue areas, cites the United States as a "special case, in respect of its enormous size and influence in the international system," calling it "by far the most important actor in most intergovernmental organizations [and] often the only actor whose participation is single-handedly decisive for the success or failure of [an international] organization" (1990, 325). This view also reflects the self-perception of the United States itself as the leader, or the "only superpower," of the international system in general (Panjabi 1997, 239; see also Paterson 1992).

In the environmental area, DeSombre (2000) focuses on US leadership in the international environmental policy arena, not only because of the United

States' economic size and influence but also because the United States has some of the most stringent environmental regulations in the world (see also Mathews 1991; Gjerde 1996; Taib 1997; Harris 2001; Sussman 2004). Haas calls the United States a "hegemonic leader" in the successful ozone negotiations.[11] Parson, who asserts in the introduction to his detailed history of the creation and evolution of the ozone regime that "international leadership, by either a single hegemonic nation or a coalition, were not important causes of the regime's formation" (2003, 8) acknowledges elsewhere that "there are some respects in which active US backing, given its size and clout, is necessary to sustain any operational global agreement" (2003, 143).

This concept of United States as leader of the international system is not to be confused with "hegemonic stability theory" (HST), popularized in the 1970s and 1980s, which asserts that success in creating regimes requires the participation of a hegemonic power (Zürn 1998, 620). In particular, the assumptions of HST—the idea that the good to be provided is a public good and that power is the factor that varies—do not coincide with the notion of leadership developed here. First, public goods are not divisible; that is, their use by one person does not deplete the amount available for others, the classic example being the lighthouse, whose light shines for all no matter how many sea craft take advantage of it. Environmental goods, however, are not public;[12] they show at least some degree of divisibility. The concept of leadership in a negotiation situation does not assume that the good provided is public, or that there is a common interest. Leadership says nothing about whether all states will ultimately benefit more under an agreement than under the current status quo; for example, it leaves open the possibility that agreement may be coerced simply by lowering no-agreement alternatives for anti-agreement states rather than by raising the value of agreement for them.

Second, I adhere to Snidal's view that "the US retains more than sufficient resources to achieve its goals" and that what varies is not its capabilities but its "quality of leadership", which itself "depends on a conjunction of resources and initiative" (Snidal 1990, 345, 346), closely matching my own breakdown of leadership into capability and will. The question of concern in this study is, therefore, what are the conditions under which the United States will be willing to lead? I propose to explain why US interest in effective agreement varies and how this is linked to successes and failures in achieving successful agreement to address international environmental issues.

An Economic Theory of Leadership

The willingness of the United States to lead negotiations to an effective agreement depends on its explicit or implicit perception of the costs of agreement versus the benefits to be obtained from agreement. Although both its will and its capability are affected by the costs of action, they differ in that will depends not just on having enough resources to undertake an action but also on deriving a benefit from the action that is greater than its cost. The assumption that decision-makers often attempt to assess the costs and benefits of different policy alternatives is well accepted in the literature (see, e.g., Most and Starr 1989; Cairncross 1992; Sprinz and Vaahtoranta 1994; Rowlands 1995; Porter, Brown, and Chasek 2000).[13]

At the initial stage, there may be as many as two or three possible benefits that an effective agreement might be expected to bring: First, of course, there is the desired environmental benefit that gives rise to the negotiations to begin with. This is calculated as the environmental damage that will be avoided as a result of the policy action, corresponding to economists' conventional measure of benefits from environmental improvement (Dudek and Oppenheimer 1986, 362). The value of this type of benefit increases with a deteriorating status quo as deterioration progressively increases the environmental costs of taking no action.

There may also be a potential benefit from the avoidance of costs associated with the loss of market competitiveness in situations in which domestic US actors are held to higher environmental standards than are actors in other countries. This type of cost is consistent with DeSombre's theory of domestic sources of international policy, except that I explicitly posit that the state of origin of the domestic policy in question matters a great deal. In the case of both ozone depletion and deforestation, the United States proposed global conventions, one reason being the fact that US laws were on the books that held US producers to high standards when compared with the rest of the world.

A third type of benefit from agreement that may be found in some cases is a positive, particularistic benefit, as opposed to simply the avoidance of economic costs. Such a benefit would include, for instance, greater profits that may be generated from the discovery and production of substitutes for regulated substances. This kind of "Stiglerian" benefit of environmental regulation, as Oye and Maxwell (1994, 28) label it, is concentrated on the few while costs are diffused across the many.[14] Environmental controls may work most effectively, according to these authors, when they confer palpable benefits upon the regulated. Oye and Maxwell identify the ozone case as an example of a Stiglerian situation, because producers of CFC substitutes benefited from regulations that mandated product substitution (1994, 595–603). It should be noted that the industry that benefits from the creation of substitutes will not necessarily be the industry that incurs the most loss from regulation of the substances that the substitutes are intended to replace. For example, there is little overlap between the fossil fuel industry and the alternative energy industry in the United States. In such a case, where industrial winners and losers from proposed international regulation are in direct conflict with each other, the discrepancy between them is accommodated by my framework as long as the position of the United States in the negotiations coincides with the economic interests of the industry that of the two has the greater effect on the US economy.

There are three types of possible costs that may be associated with any effective global environmental agreement. In any agreement there is a potential cost of halting activities banned or regulated by the agreement and those that depend on the regulated or banned activity or product. Those countries for which an effective agreement on a particular environmental issue poses no cost are more likely to support it. For instance, Australia put forth an idea for a South Pacific Whale Sanctuary in the 1990s. Because the Australian whaling industry had died out earlier there was no cost to as it would not be affected economically at all, but such a pro-environmental stance earned Australia a few intangible benefits in the form of "good greenie points".[15] Second, and alternatively, if substitution is possible for any regulated activity or product, there is the cost of research, development, and marketing of such substitutes. Finally, there are the potential costs of manipulating preferences to bring other states into an agreement, given the preference of some states for the status quo.

The United States' cost-benefit calculation can shift; this can happen either because of a change in circumstances or because of an assertion of what Sebenius has called "value-claiming issue linkage".[16] Scientific discoveries or new interpretations may increase the amount of benefit expected in the form of avoidance of environmental damage. Technological discoveries can contribute to potential positive economic gain from new substitutes for regulated activities or substances, while at the same time decreasing the costs to be incurred from halting any affected activities. On the other hand, if the United States incurs unexpected unrelated costs, this increases the strain on the Treasury, which can change perceptions of the cost of undertaking effective environmental action. One illustration of this type of phenomenon may be found in the US Administration's changing perception of the costs of levee repair around New Orleans as it undertook to provide tax cuts, restructure security within the United States and wage a costly war in Iraq in 2003. Between 2001 and 2004 the amount of federal money spent on Southeast Louisiana urban flood control halved, from a high of $69 million in 2001 to $34.4 million in 2004.[17]

Value-claiming issue linkage can increase the cost of preference manipulation. This takes place when one party to a negotiation conditions its cooperation in a negotiation on concessions in other areas, such as financial or technological assistance. The cost of such concessions may make agreement too expensive for the pro-agreement side in comparison to the benefits it is expected to bring. One may assume that some costs of manipulation are built into initial calculations of the costs of agreement for the United States. Nevertheless, value-claiming may push the prospective costs of manipulation up so much that the cost-benefit equation for the United States shifts, even in situations in which US calculations otherwise favor an effective agreement. Take, as an example, the case of efforts to agree an international convention to halt the loss of biological diversity—another environmental issue, along with climate change and deforestation, that was addressed through negotiations for a binding global convention in the period leading up to the 1992 UN Conference on Environment and Development (UNCED) in Rio de Janeiro. In that case, value-claiming issue linkage in the form of developing countries' demands for concessions on intellectual property rights and transfer of biotechnology threatened the entire negotiation process and ultimately caused the United States to refuse to sign the Convention on Biological Diversity when it was opened for signature.

This raises the question of why, given its size and power resources, the United States cannot prevent such value-claiming. Action to forestall value-claiming demands must have the effect of lowering the value of no agreement for that side. A party attempting to claim value must be able to make a credible threat to walk away from any agreement that does not meet its minimum demands—in other words, it must be willing to keep the current status quo or to make an agreement not very far removed from the status quo. A negotiating party has the ability to walk away from the agreement only if the net value to it of agreement is lower than the value of its no-agreement outcome. If the value of no agreement can be manipulated such that the value of no agreement for that party becomes lower than the value of agreement, it will no longer have the will to walk away from agreement, and with that its insistence that its demands be met will also be greatly weakened. It can bluff, of course, but then its credibility will come into question, depending on how much the other side knows about its true valuations.

In effect, this means that if the pro-agreement side can make credible threats of punishment, it can avoid the cost of incentives. Its motivation to do this will depend on the cost of potential punishment and the risk of actually having to take such action, versus the cost of providing incentives. The ability to influence another party through the use of credible threats rather than incentives is of course preferable cost-wise to the coercing party. If a threat is successful no cost will be incurred because the threatened party will be coerced into specific behavior in order to avoid the threatened action.

However, the ability to make a credible threat will be influenced by the context in which it is made. For instance, the United States may use threats of trade restrictions—assuming a US market for relevant goods from relevant countries and US willingness to use unilateral trade restrictions. Its willingness to use threats may be affected by the potential effect of restrictions on economically powerful domestic actors and by international trade rules. Other hypothetical threats it could use include threats to existing aid programs, for relevant countries, or even a military threat in an extreme case. A military threat would probably lack credibility, however, because of the risk of severe costs it could bring. It would not be credible, for example, for the United States to threaten to attack a country that opposes a treaty to halt deforestation simply in order to lower the value of no agreement for that country.

Because of the need for specific conditions to exist in order to make threats credible, in a situation of value claiming where credible threats do not exist, the value of agreement relative to the costs a pro coalition expects to incur in meeting value-claiming demands or providing other incentives will determine whether the coalition will be willing to bear such costs. If it is not, it may attempt to obfuscate language on obligations while still striving for some agreement, such as through the insertion of "waffle" language[18] that qualifies proposed obligations on such demands, frequently to the point of their elimination altogether. This, indeed, can be said to be the case with the Convention on Biological Diversity (CBD). As Humphreys points out, language appears in numerous Articles that qualifies the commitments states make under the Convention, such as "as far as possible and appropriate", "where feasible", "where necessary", and "in accordance with national legislation and policies" (2005, 3, fn 18). In extreme cases, an otherwise pro-agreement country may choose to opt out of agreement altogether, as, again, was the case of the United States with regard to the CBD.

There are a number of difficulties in attempting to calculate costs and benefits quantitatively. These include difficulties in identifying all relevant costs and benefits, identifying transfer payments (where a gain to one person is cancelled out by a loss to another), pricing benefits and costs not evaluated by the market mechanism, and determining whether existing prices represent the true (social) value to society of using those resources in a particular way. With specific regard to environment and natural resource issues, there are difficulties in comparing long-term benefits with short-term costs, given the convention of discounting the future and particularly in light of the prospect that some uses of environmental and natural resources may bring about negative long-term changes that are irreversible (Eilperin 2006, A01). Differences between countries and between subject areas affect discount rates, too. For example, the US government uses a social discount rate of 7 percent as its basic guideline for calculating costs in some areas of activity, but for long-lived projects with impacts over many generations the federal government issued guidelines in 1995 allowing for the use of lower discount rates (Stiglitz 2000, 287–288). However, a 10 percent or even 11 percent rate has also been used historically (Dudek and Oppenheimer 1986; see Cooper 1992).

It is not my purpose here to advocate any particular method of valuing long-term environmental benefits, nor to perform independent cost-benefit analyses of the cases in order to determine whether the United States' position on any disputed area conforms to an independent assessment of costs and benefits. Rather, I consider as closely as possible the valuations used by the US federal government in its considerations of the cost of action versus inaction—and the cost of agreement versus no agreement—in the cases examined here. The judgments of US officials themselves on the net value to the United States of the total stream of benefits and costs, including transaction costs, at the times in question are taken as the basis for analyzing their actions on international policy in each case considered here.

Where explicit cost-benefit analyses have not been performed, or where evidence linking the formation of US government positions explicitly to any formal cost-benefit analysis (CBA) is lacking, we can assume at least implicit consideration of questions such as the following; the answers to these questions can help to establish the parameters of an implicit cost-benefit assessment in order to explain the level of US interest in effective agreement in a particular issue area:

- Does a proposal for agreement call for domestic action on the part of the United States (or US actors)? If not, then the cost of the proposal is low for the United States, in terms of the cost of developing substitutes or limiting economic activity.
- Does a proposal for agreement require action on the part of other countries? If so, then the proposal holds the potential to contribute to the benefit side of the equation for the United States, in terms of improving the environment and/or raising standards in other countries that will have the effect of lowering market competition.
- If the proposal requires action on the part of the United States, does it bring with it the promise of positive benefits for the United States, such as the creation of new markets or a competitive advantage in new technologies? If so, then these potential gains would contribute to the benefit side of the cost-benefit equation.
- How much will it cost to coerce agreement by opposing countries? This cost will add to the cost side of the equation.

While such questions may apply to many countries, the answers to them are more likely to determine valuations for the United States than for other countries because while the possibility exists that the United States can manipulate the valuations of others, it is far less frequently the case that its United States' own preferences can be manipulated by other countries toward favoring agreement.

The remainder of this book is devoted to a detailed examination of three negotiation histories in support of the above assertions. We shall see that even within the success story of the ozone regime, the United States' cost-benefit structure led to effective regulation to restrict emissions of some ozone-depleting chemicals but less restrictive regulations on others. We will also be confronted with the paradox that the United States was the first state to propose a stand-alone forest convention in 1990, yet was unable to bring it to effective completion thanks to an unfavorable cost-benefit calculation. Finally, we will grapple with the costs and benefits to the United States of international regulation on greenhouse gas emissions, and propose new strategies for achieving US cooperation on effective international regulation based on the explanatory framework developed here.

CHAPTER 3

Ozone Politics

Efforts to protect stratospheric ozone[1] from depletion due to anthropogenic causes have succeeded beyond what some ever thought possible when the harm was originally discovered, and have done so in ways that have set precedents for international environmental diplomacy in other issue areas (Lang 1997; Benedick 1998;[2] Downie 1999). The ozone regime, comprising the 1985 Vienna Convention for the Protection of the Ozone Layer, the 1987 Montreal Protocol (MP) on Substances that Deplete the Ozone Layer, together with various amendments and adjustments to the Protocol that have been agreed since 1990, continues to be the most successful environmental treaty system currently in force. These treaties will be referred to collectively here as "the ozone agreements". True, the condition of the ozone layer, particularly over the Antarctic, is expected to worsen over the next few years before it begins to improve consistently, but if fully implemented, scientists predict that the ozone agreements will lead to full recovery of the ozone layer by between 2050 and 2060 (Reuters 2005; UNDP 2005b; Adam 2006).

Because of their success, the ozone agreements have proven to be an extremely popular case study for academics, with many explanations offered for their effectiveness and "lessons" drawn for other environmental issue areas. Indeed, of the studies of international environmental cooperation written in the 1990s, a huge proportion focuses on the ozone negotiations. Among these, some studies approach stratospheric ozone as a CPR (common pool resource), while others argue that it is not. Other writers delve somewhat into differences in interests: Peter Haas (1992a) puts forward an explanation based on the existence of an epistemic scientific community, while Litfin (1994) argues that science itself has been at the service of other interests. "Science versus economics" has been a theme in studies of the ozone case such as those of Downie (1996), Sprinz and Vaahtoranta (1994), and Oye and Maxwell (1994). There are many other attempted explanations, but certainly the most comprehensive picture of the ozone regime to date is given by Edward Parson in his 2003 study of the ozone regime. His work brings science, technology, and policy all together into a multifaceted inductive explanation of this exceptionally successful case, and then, once again, attempts to draw lessons from these for the management of other issues.

While the present work cannot—and does not attempt to—match the breadth of Parson's on this particular case, the explanation offered here is, I believe, broadly compatible with his. In addition, it accomplishes two further aims: First, use of the simple theoretical framework laid out in chapter 2 provides a structure into which to pin the elements that are most important to the success of the ozone case. Second, limiting this case study to those areas of debate that were most contentious and to the factors that are most explanatory leaves room for analyzing the outcomes of two other cases along the same lines in order to test the explanatory value and generalizability of my proposed framework.

My framework is also in some ways more comprehensive than Parson's, in that it incorporates both the science and economics of ozone management as well as the role of power and influence. This chapter starts from the argument that US leadership is necessary for the achievement of an effective agreement in a situation of asymmetric deadlock, leadership being defined as willingness to use coercion to shift the preferences of other states through raising the benefits of effective agreement for them or raising the costs of nonagreement. After providing some background on the problem of ozone depletion, I examine the effectiveness of agreements on chlorofluorocarbons (CFCs) according to the criteria of effectiveness laid out in chapter 2 and describe the United States' role in achieving an effective agreement to phase out CFCs through the Montreal Protocol. Finally, I show how the typology described in chapter 2 explains the United States' willingness to manipulate international cooperation into an effective regime to halt and reverse stratospheric ozone depletion. In chapter 4 the examination of the ozone agreements continues with a look at less frequently discussed disputes over hydrochlorofluorocarbons (HCFCs), methyl chloroform (MC), carbon tetrachloride (CT), and methyl bromide (MB).

The Problem

The stratospheric ozone layer performs a critical environmental function of absorbing ultraviolet rays that are harmful to most life forms on earth. Ozone (O_3) is a molecule formed from three oxygen atoms, compared with the two atoms in breathable molecular oxygen (O_2). Oxygen molecules break up in the powerful sunlight of the thin stratosphere and re-form into ozone, which is then broken down again by the same sunlight into oxygen and monatomic oxygen in an endless cycle. Through this process the ozone intercepts and filters shortwave ultraviolet radiation (UV-B), thus UV-B cannot penetrate living tissues.

Ninety percent of lower atmospheric ozone is found in the stratosphere. Its natural distribution varies according to latitude and season but overall the concentration of ozone in the stratosphere is very small, about 10 parts per million (ppm). Current knowledge indicates that it has been decreased significantly by anthropogenic activity, in particular the release of CFCs into the atmosphere. As ozone diminishes, the amount of UV-B reaching the earth's surface increases. The risks posed by this are thought to include higher incidence of skin cancer and suppression of the human immune system, cataracts and infectious diseases in humans and animal life, genetic mutations and lower productivity of plants, disruption of the aquatic food chain, increased climate change, and deterioration of synthetics and plastics. Other potential consequences, such as changes in the stratosphere's chemical composition, are unknown (Thomas 1992).

Fears about the effects of supersonic transport on the ozone layer were first expressed in the United States in 1970. Supersonic transport was abandoned in the United States for economic reasons, but in 1974 two studies appeared that together provided a different cause for concern. Richard Stolarski and Ralph Cicerone, two atmospheric scientists, published a study showing how chlorine in the stratosphere could destroy ozone through a catalytic chain reaction. Stolarski and Cicerone were mainly concerned with chlorine associated with space travel; it was a pair of chemists, F. Sherwood Rowland and Mario Molina, who linked this phenomenon to CFCs. They found that CFCs are peculiar in that they are not broken down quickly in the lower atmosphere but are so chemically stable that they persist and rise slowly into the stratosphere. They can remain intact for years but eventually they are broken down by solar radiation, releasing vast quantities of chlorine into the stratosphere. Together, these two studies showed the first connection between the use of CFCs and depletion of stratospheric ozone.

The World Meteorological Organization (WMO) issued the first authoritative statement that stratospheric ozone was being depleted in 1975[3] and global atmospheric models in the 1970s predicted a slow downward trend in stratospheric ozone levels. Although the Rowland-Molina hypothesis on the connection to CFCs was not confirmed until 1988, policy-makers in numerous industrialized countries began to see ozone depletion as a global problem and CFC use as perhaps a major cause of it. The United States and eight other countries banned the use of CFCs in aerosols between 1975 and 1980 and called for a global ban. Over the course of the next 25 years, the international community produced the 1985 Vienna Convention, the 1987 Montreal Protocol, four amendments to the Protocol that were adopted at various Meetings of the Parties to the Protocol (MOPs), and several significant "adjustments". There has been a total of six separate ratification processes associated with this set of agreements so far; the adjustments are a mechanism unique to the Montreal Protocol for strengthening commitments on already controlled substances without requiring ratification.

The Ozone Agreements and the Criteria of Effectiveness

Insufficient time has passed for the world to know now whether the international commitments made with regard to ozone depletion are adequate. Yet the effectiveness of agreements to halt the consumption and production of CFCs can readily be discerned even without the passage of time, through the criteria of effectiveness laid out in chapter 2.

Commitments: This criterion measures how far the agreements, taken as a whole, actually commit the parties to take action. The ozone agreements contain commitments on precise timetables and targets pertaining to different chemicals covered. Although the original 1987 commitments under the Montreal Protocol covered only five CFCs and three halons, commitments to control other substances have progressively been added through various amendments, and existing controls tightened through adjustments made at subsequent MOPs.[4] As each new chemical came under scrutiny, comparisons of their deleterious effects were enabled through the creation of a scale to measure their "ozone-depleting potential" (ODP). The ODP number relates the harmfulness of a CFC or related compound, weight for weight, to the harmfulness of CFC-11, which is assigned the ODP score of 1. It consists of a rising scale for how many chlorine or bromine (even worse than chlorine) atoms the compound in question has and an allowance for how long the compound stays in the atmosphere (Bellany 1997, 153).

It is overall cuts in calculated production and calculated consumption that are sought. The calculated level of production of a controlled substance equals the actual amount produced multiplied by its ODP. Total calculated production is the sum of the calculated production of each substance minus the amount destroyed and minus the amount used as feedstock. Consumption is defined using a surrogate of production plus imports minus exports of controlled substances; the calculated level of consumption for a controlled substance is determined by the actual amount of consumption multiplied by its ODP (UNEP 2000c).

By far the most significant commitments of the ozone agreements pertain to CFCs, specifically CFC -11, -12, -113, -114, and -115, which consist of precise targets for phase-out of production with rapid timetables for their achievement, thus ranking very high in effectiveness. Beginning with a baseline cap on consumption, based on 1986 levels, countries were required to freeze consumption of these in 1989 (MP), cut consumption by 75 percent by 1994, and phase it out by 1996 (MOP-4, Copenhagen, 1992). "Article 5 countries" (in other words, developing countries) were required to freeze consumption in 1999, cut consumption by 50 percent by 2005 and by 85 percent by 2007, then phase it out by 2010 (MOP-7, Vienna, 1995).[5]

Compliance: The dispute settlement procedure is laid out in the Vienna Convention. The options for resolution include negotiation, mediation, or arbitration, or submission to the International Court of Justice (ICJ). If both parties to the dispute cannot agree on one of these, one party can request the creation of a conciliation commission. Such a commission would be composed of an equal number of members appointed by each party concerned, which "shall render a final and recommendatory award, which the parties shall consider in good faith" (Article 11). None of these options is compulsory unless a party accepts arbitration or ICJ submission as compulsory.

The lack of a compulsory dispute settlement procedure is offset by mechanisms that were later laid out under the Montreal Protocol. These mechanisms are intended to create an enabling environment for carrying out the substantive commitments of the Protocol. They have evolved over the life of the Montreal Protocol and include agreements on noncompliance mechanisms, restrictions on trade in controlled substances, and a fund to assist developing countries with implementation of commitments.

With regard to monitoring, parties are responsible for reporting statistical data to the secretariat on their annual production, imports and exports of each controlled substance to parties and non-parties (Article 7[6]). Data submitted is not reviewed by any third party or entity (Benedick 1998, 273); however, under the Noncompliance Procedure adopted at MOP-4 (UNEP 1992) and Article 8, parties themselves, or parties suspecting noncompliance, or the secretariat can request the Implementation Committee established by the Noncompliance Procedure to address a suspected or possible situation of noncompliance. Nevertheless, there is no generic identification of situations that could constitute cases of noncompliance against which to check the actions of parties suspected of noncompliance, despite a MOP-3 Decision requesting this.

The Implementation Committee consists of ten parties, based on equitable geographical distribution and elected for two years by the MOP. Its main function is to review implementation by considering and reporting on any submissions received by any party or the secretariat with regard to possible noncompliance. The Committee has no decision-making power, only the power to report to the MOP and give optional recommendations on how a case of noncompliance should be handled.

There are no enforcement or sanctioning procedures specified in the Protocol or in the Noncompliance Procedure. However, MOP-4 did provide an "indicative list of measures" that the MOP, as the Protocol's decision-making body, could opt to take in cases of noncompliance. These include (1) "appropriate assistance" for implementation, which could include help in data collection, training, technical assistance, financial aid, and technology transfer;[7] (2) "issuing cautions"; and (3) "suspension . . . of specific rights and privileges under the Protocol" (UNEP 1992). Such rights include the right to trade with other parties in controlled substances and products containing them (Article 4).[8] In addition, a MOP-6 Decision gives developing countries the right to have the status of an "Article 5 country"; suspension of this right would entail loss of eligibility for financial assistance if data are persistently not forthcoming.

With regard to implementation assistance, there are two facilities in the ozone agreements that take account of countries unable to comply with timetables for reduction and phase-out of ozone-depleting substances (ODS). First, Article 5 of the Protocol explicitly puts developing countries whose annual calculated level of consumption of a controlled substance is less than 0.3 kg per capita into a special category, enabling them to take advantage of delayed compliance (and, since MOP-2, financial assistance). Countries that fall into this category are labeled "Article 5 countries." In its present form, as modified by successive MOPs, Article 5 specifies different periods of delay for different chemicals, based partly on when the chemical came under controls and partly on negotiated compromises. In addition, it gives any Article 5 country the right to notify the secretariat that it is unable to implement any or all of its obligations on controls due to inadequate implementation of financial and technical assistance commitments by developed countries as laid out in Articles 10 and 10A agreed at MOP-2. The parties are required to consider the matter at their next meeting and decide upon appropriate action.

The second facility to assist implementation is the Multilateral Fund for the Implementation of the Montreal Protocol (MLF) within Article 10, also established at MOP-2. Directed explicitly at "[P]arties operating under paragraph 1 of Article 5," the MLF constitutes a major undertaking toward compliance assistance as its purpose is to meet all "agreed incremental costs of such parties in order to enable their compliance with the control measures of the Protocol." The MLF is financed from "assessed" contributions by non-Article 5 parties (UNEP 1990). An Executive Committee of 14 party members, rotated within seven developed country regions and seven developing country regions, formulates overall MLF policies under the MOP's authority, with the World Bank, the United Nations Environment Programme (UNEP) and the United Nations Development Programme (UNDP) dividing up operational duties.

Participation: The question of participation pertains to the rate and quality of participation in terms of ratifications of the Vienna Convention, the Montreal Protocol, and the four Amendments to the Protocol that have separate ratification requirements. No amendment may be ratified without ratification of all instruments preceding it. As of May 25, 2006, participation in the Convention and Protocol was near-universal, with 190 and 189 state parties, respectively (UNEP 2005b). This does not include the various Amendments to the Protocol, which deal with other chemicals.

Producers of the problem are generally developed countries, thus this aspect of participation is generally well covered in those four first agreements (the Vienna Convention, the Montreal Protocol, the London Amendment of 1990, and the Copenhagen Amendment of 1992) that were signed by all developed countries. Major potential developing country producers of the chemicals covered are also well represented in these four agreements.

The potential contributors of financial and technical resources to assist developing country implementation are all parties to the first four agreements. Of the countries that have not ratified the London Amendment establishing the financial mechanism, and the Copenhagen Amendment that succeeded it, all but a handful are developing countries that would presumably be eligible for assistance under Article 5. The rest are so-called "countries with economies in transition," in other words, Eastern European countries.

Relationship to other treaties: None of the ozone agreements contain language regarding their relationship to other treaties, either in general or on any of the specific points laid out in my criteria of effectiveness.

Table 3.1 provides an overview of how well the ozone agreements have met the criteria of effectiveness according to scores on several subcriteria within each area. As can be seen, the ozone agreements do not score perfectly on the criteria of effectiveness. In particular, enforcement and relationship of the ozone treaties to other international treaties are not explicitly delineated, although arguably enforcement procedures may not be necessary for implementation if other procedures and mechanisms work well enough to assure a sufficient level of compliance. What is certain is that the commitments made in these agreements are focused and specific, and timetables and methods for achieving compliance are detailed. The willingness of states to acquiesce to this international authority is also seen, in the near universality of

Table 3.1 The Ozone Agreements and the Criteria of Effectiveness

Commitments	*Compliance*	*Participation*	*Treaty relationship*
precise targets	specified dispute resolution body	universal participation[a]	language specifying relationship to other instruments
2	1	2	0
timetables for their achievement	specified monitoring or review procedures	producers of the problem	superiority to other instruments
2	1	2	0
firm, concrete commitments rather than "indicative goals"	body established to review implementation	potential providers of financial resources to assist implementation	specified procedure for resolving conflicts with other instruments
2	2	2	0
	enforcement procedures/sanctions for non-compliance	potential providers of technology to assist implementation	specified body to resolve conflicts with other instruments
	1–2	2	0
	mechanism for assisting implementation		
	2		

Notes:
[a] Because the ozone regime consists of separate agreements and amendments, each must be ratified separately. The number of ratifications relevant for CFC controls, as of July 18, 2006, stands at 190 for the 1985 Vienna Convention, 189 for the 1987 Montreal Protocol, 183 for the 1990 London Amendment, and 174 for the Copenhagen Amendment (UNEP 2005).
Coding:
0 = nonexistent
1 = somewhat
2 = effective

ratifications of the Protocol itself and the sizable and ever-growing number of ratifications of the Amendments to it. Therefore, this set of agreements rates very high in effectiveness, particularly on commitments on halting CFC emissions.

The Story of CFC Regulation

The most hard-fought disputes in the ozone negotiations have centered around commitments to restrict and phase out production and use of ozone-depleting chemicals, the first and most important indicator of effectiveness. Most of the controversy that has been engendered within ozone negotiations, either within the developed world or between developed and developing countries, has centered around five ozone-depleting chemicals; of these,

CFCs were the first to be regulated internationally and were also perhaps the most hotly debated.

Chlorofluorocarbons are identified as the leading cause of stratospheric ozone depletion because of both the amount of CFCs produced since the 1930s and the amount of ozone-damaging chlorine emitted per unit of the chemical. However, it took a full decade to go from calls for an international ban on CFCs to agreement on the first international controls on CFCs under the 1987 Montreal Protocol. These controls consisted of a commitment among developed countries to reduce consumption of certain CFCs by 50 percent by the year 2000. Only three years later, in the 1990 London adjustments, those countries committed to phasing out all CFCs by 2000, and in 1992 they accelerated this phase-out to 1996. The development of CFC controls closely tracks the evolution of perceptions of costs and benefits in the United States that resulted particularly from growing knowledge of the environmental costs of inaction combined with technological developments that could lead to positive, particularistic economic benefits for the US.

The United States and CFCs

The United States was an early proponent of international cooperation to halt ozone depletion, voicing the need for an international mechanism to control CFC in aerosol uses as early as 1976.[9]

The United States was prepared to be a leader on international cooperation to halt ozone depletion, both in the process of defining the issue and in pushing for international action. This despite the fact that it was the largest global producer of CFCs, the primary ozone-depleting gas. Indeed, the United States was the only major CFC producer to call for international regulation.[10] American leadership during the early stages, however, corresponded to an initial set of factors that gave it an interest only in global regulation of CFCs in nonessential aerosols.

Chlorofluorocarbons became an issue of public concern in the United States in the 1970s. As a result of consumer reactions to revelations of potential dangers to the ozone layer starting in the early 1970s, the US market for spray cans containing CFCs fell by two-thirds by 1978 (Benedick 1998, 28). By 1977, legislation against CFCs in spray cans had been passed in Oregon and New York and was also introduced in other states and municipalities (Dotto and Schiff 1978; Parson 2003). The 1976 Toxic Substances Control Act gave the Environmental Protection Agency (EPA) broad regulatory authority over CFCs and by 1978 had been used to prohibit the nonessential use of CFCs in aerosols nationwide (Morrisette

1989). The US Food and Drug Administration (FDA) defined "essential" to mean that (1) there were no technically feasible alternatives to the use of CFCs in the product; (2) the product provided a substantial health, environmental, or public benefit unobtainable without the use of CFCs; and (3) the use did not involve significant releases of CFCs to the atmosphere, or the release was warranted by the benefit conveyed. Other products in which CFCs were the *active* ingredient rather than the propellant were not covered under the ban (Dotto and Schiff 1978).

Some analysts highlight public perceptions of environmental and health risks (Morrisette) or the media's success in "capturing public imagination" as the main reason behind the US legislation (e.g., Weiss 1993). Others focus on its enactment despite early industry opposition; for instance, DuPont declared in 1975 that restrictions on CFCs "would cause tremendous economic dislocation" (Benedick 1998, 31). However, the economics of the situation are straightforward, clearly favoring a federal ban on nonessential CFC use in aerosols.

First, although the effects of ozone layer depletion were yet to be studied, the National Academy of Sciences reported in 1976 that eventual ozone layer depletion could reach 7 percent. This represented a threat of significant costs from inaction even if the costs could not yet be calculated. The cost of action was great—the EPA estimated in 1977 that action in the form of switching to other propellants and packages would cost the industry $169 million to $267 million per year for four years after the ban. However, there was the possibility of a *savings* of $58 million to $240 million from cheaper alternatives (Dotto and Schiff 1978, 288). Producers were already moving to develop new propellants for spray cans before the 1978 ban was introduced. Dotto and Schiff report that the head of one aerosol company announced that they had developed a new aerosol system one day after the FDA's timetable for regulations banning CFCs was released. Indeed, only half of the aerosol products in the United States were propelled by CFCs even before consumer pressure against them mounted (Rowlands 1995). Some substitutes, such as hydrocarbons, were actually cheaper than CFCs and had already been used for many years in some products. There was therefore a $165 million saving for the US economy in 1983 alone from the substitution of hydrocarbons for CFCs in aerosols (Shea 1988, 24). On the down side, some substitute propellants also released ozone-depleting chemicals into the atmosphere (Roan 1989, 203) whereas others, such as hydrocarbons, would later be implicated in global warming. However, given that aerosol use of CFCs was growing fast, even though most aerosol uses were indeed nonessential and substitutes did exist for most such uses, banning them would be a quick fix for part of the emissions problem. This made the choice between maintaining CFCs in aerosols and halting their use in light of the potential risks relatively easy for regulators (Parson 2003, 41).

There were other considerations that were favorable to the decision to ban CFCs in nonessential aerosols. Consumer sales of aerosol cans had already plummeted 25 percent in 1975, after publicity about the scientific reports showing a connection between ozone loss and CFCs, and by the time the EPA announced its draft regulations in 1978 consumer use of aerosol CFCs was down by 75 percent (Parson 2003, 40). Controls on CFCs in aerosols only threatened a small fraction of the chemical industry's business and were not worth falling on one's own sword for, particularly for companies that could easily reformulate their products or whose main product line did not depend on CFCs (Parson 2003, 36–37). Marketers who were not solely dependent on CFC-containing aerosol products also benefited from good publicity by immediately exploiting the ozone crisis to market their non-CFC aerosols and non-aerosol sprays as "ozone-friendly" in order to drive up sales (Dotto and Schiff 1978, 164). For other producers, the national ban provided an excuse to find substitute aerosol propellants that were easy and more economical to produce and benefited them by lending uniformity to the patchwork of varying state regulations they were beginning to face. It is true that some members of the industry suffered; producers looked for other markets for CFCs to offset the loss of aerosol markets, and those markets did grow, but they could not make up for the loss of the largest CFC market (Parson 2003, 113).

All in all, however, it was a case of net economic benefit from taking action relative to no action. Benedick confirms this: "In effect, the [United States and other governments] determined that the benefits of CFC-propelled sprays were negligible when weighed against the potentially huge costs if the scientific theories proved correct" (1998, 24). Bellany (1997, 148) notes that because the halt on US production of CFCs for aerosols immediately produced a drop of more than one-third of all CFC releases into the atmosphere, the benefits could be seen as heavily outweighing the short-term costs to US manufacturers of producing new aerosols, even given that these benefits of CFC reduction could not be captured by the United States alone.

These actions in the United States, however, were based on only half of the legislation actually passed in the 1970s. Aerosols were only half of the problem, and it was always intended that non-aerosol uses of CFCs would be regulated next (Parson 2003, 55). An amendment to the US Clean Air Act, passed in 1977, gave an even broader mandate than the Toxic Substances Control Act, authorizing—indeed, requiring—the EPA administrator to regulate any substance that might be reasonably anticipated to affect ozone in the stratosphere if it might endanger public health or welfare. It thus set a precedent in lowering the threshold for governmental measures to protect the ozone layer by not requiring scientific certainty (Benedick 1998) and not requiring an explicit comparison of benefits and costs (Parson 2003). This legislation was intended to cover other uses of CFCs, and in fact an EPA-led multi-agency work group, formed to facilitate the rulemaking process on CFCs, originally proposed a two-phase effort that would regulate first nonessential uses of CFCs and then other uses.

The second-phase EPA proposal for freezing nonaerosol uses under the Clean Air Act was put forward in 1980, but despite the mandate of the legislation and the work group, this proposal never got off the ground. Consumer response to the EPA proposal was negative this time, and the multi-agency group itself argued that it would not be "viable", due to little interest and cooperation in regulating CFCs outside the United States.[11] Numerous authors, however, point to the US chemical industry's resistance to freezing nonaerosol uses unilaterally as influencing the decision not to proceed. The industry formed the Alliance for Responsible CFC Policy (ARCFCP) to lobby against regulations. The ARCFCP persuasively argued for time to develop substitutes for other uses and for a "level playing field" with regard to other countries (Morrisette 1989; Roan 1989; Benedick 1998).

On a broader level, it was clear to US regulators that a ban on all CFCs would not be cost effective. A government study performed by the RAND Corporation calculated that the prospects for substantive technological advancements in substitutes for a variety of uses were not bright and that a ban on all CFCs from 1978 could cost billions of dollars and adversely affect the jobs of millions of workers.[12] Although the aerosol ban itself did not ultimately cause huge job losses, the government listened to industry's predictions regarding all CFC uses. Indeed, the conclusions of the RAND study were themselves weighted toward the industry's own positions, given that the researchers had to rely on the CFC industry for data and were limited in their ability to verify its accuracy (Parson 2003, 92; Metzger 2005). Consumer response this time did not push industry toward regulation as it had with nonessential aerosols. Perhaps due to complacency, consumer attention to the ozone crisis in general declined sharply after the initial regulations had been achieved (Parson 2003, 42). What reaction there was was unexpectedly negative, thanks to a successful publicity campaign by the ARCFCP. The EPA received thousands of negative comments—more than 2,500 from DuPont customers alone—in response to its proposal to regulate nonaerosol uses of CFCs (Brown and Lyon 1992, 131). All this despite the fact that the aerosol ban was in fact accelerating the growth of CFCs in other uses, because manufacturers needed other markets for their excess capacity (Parson 2003, 42).

The perceived future benefits from further regulation did not balance these higher costs, for two reasons: First, scientific calculations of ozone depletion based on then-current models were gradually being revised downward, thus estimates of negative effects on health were also falling. Second, because of the regulations already put into effect there were fewer CFCs being discharged into the atmosphere and thus the level of eventual ozone depletion was thought to be decreasing, again decreasing the magnitude of perceived benefits. Spray cans represented about 75 percent of the problem in the United States (Dotto and Schiff 1978), so any benefit to be gained from further action against CFCs would be much less than the benefit that had been gained from banning aerosol uses.

Finally, Rowlands associates the election of Reagan in 1980 with an effective rise in discount rates as emphasis was given to short-term individualistic benefits at the expense of future benefits. An executive order signed by Reagan that required a cost-benefit analysis (CBA) to be made for all new major regulations manifested this new attitude. Data on the cost to an economy of not taking an action are usually not available and such costs are not easily calculated. There were also no studies demonstrating how the damage attributable to ozone-depleting chemicals compared with the costs of limiting their use.

As early as 1977, attempts to perform CBAs on the costs and benefits of banning CFCs had been made, but Environmental Resources, a British NGO, reviewed several such studies at the original 1977 UNEP meeting and concluded that there were numerous problems with taking such an approach, particularly at the global level and particularly for environmental and natural resource issues. The group concluded that a conventional CBA in this area was not entirely appropriate, although they listed qualitatively the costs and benefits of a global ban on production and distribution of CFCs (Environmental Resources Ltd. 1979). Nevertheless, such an approach was required by the US government. Given the limitations in calculations of true costs and benefits, a consideration of short-term costs easily overcame the perceived long-term benefits of regulation. Economics therefore ensured that US support for international regulation would be limited initially to a call for a ban on the use of CFCs in nonessential aerosols.

Toward International CFC Regulation: Stage One

Although the United States and others pushed for action on aerosol CFCs at the international level from the mid-1970s, it was Sweden that actually presented a resolution at UNEP's 1981 Governing Council meeting, asking UNEP to begin negotiations for an international convention to protect the ozone layer (Parson 2003, 112). The *Ad Hoc* Working Group of Legal and Technical Experts for the Elaboration of a Global Framework Convention for the Protection of the Ozone Layer began its work in 1982 (Porter, Brown, and Chasek 2000, 88). By this time, the United States' position had shifted and, while it expressed support initially it then backed off, at least for a while. Why?

First, contrary to DeSombre's assertion, the evidence suggests that US industry was not behind the push for international regulation. After all, it was CFC *use* that was banned in the United States, and only in nonessential aerosols. This meant that American CFC producers could switch their manufacturing to other uses (Parson 2003, 42) as well as continue to export CFCs. Nor did they have to face imports of CFCs in nonessential aerosols from foreign competitors. These facts, combined with the factors that inhibited united industry opposition to domestic controls in the first place, meant that there was little need to call for international regulation to "level their playing field."[13] Indeed, industry officials argued the reverse: because all governments had access to the same scientific information, the fact that other producing countries had not emulated the United States' aerosol ban was an indication that the science did not warrant an international regulatory response (Parson 2003, 59). It was not until 1985 that industry bodies attended the UNEP Working Group negotiations as observers (Parson 2003, 119).

Instead, the US EPA led the drive for international action during the 1970s, based on the scientific assessments coming out of the United States. The most pressing issue was the recognition that the problem of ozone depletion was not going to be solved by one country alone, given that there were a dozen countries producing CFCs and that the ozone layer was indeed a global commons, with ozone depletion potentially affecting all. Throughout the early years of domestic debate in the United States, US officials repeatedly called for other countries to enact CFC aerosol controls (Parson 2003, 44), and it was the environmental benefit of protecting the ozone layer that was uppermost in this drive. Prospective environmental benefits proved to be inadequate for achieving global regulation, or indeed, for sustaining US interest.

Second, the United States' preference, at least at first, was to pursue regulatory coordination at the international level in an *ad hoc* process, given limitations in existing intergovernmental organizations (Parson 2003, 45–46). By 1980, this process had failed because it gave other countries little or no incentive for action, but the United States failed to respond with support for a more forceful initiative because Ronald Reagan became president and US policy shifted, temporarily. The then-administrator of the EPA, Anne Gorsuch, a firm adherent of Reaganesque anti-regulatory ideology, did not support Sweden's proposal for international ozone regulation, which went further than US domestic controls, nor did she even advocate international regulations based on the United States' own domestic regulations. In fact, one participant at the first meeting of the *Ad Hoc* Working Group reports that US delegation members even said informally that they should not have banned aerosols domestically in 1977.[14]

Ultimately, however, economics won out over ideology shortly after the *Ad Hoc* Working Group's establishment as Gorsuch's 1983 replacement, William Ruckleshaus, recognized that it did not make sense for the United States "not to support internationally what it had already implemented domestically" (Benedick 1998, 42). This also coincided with a move of the CFC issue, after some EPA in-fighting, from the Toxics Office to the Air Division of EPA, which had more expertise and more authority over the issue under the Clean Air Act (Parson 2003, 116). Thus in 1983, the United States returned to its former position advocating an international ban on aerosol CFCs, and actually went further in some ways (Parson 2003, 116). With this move, the United States joined Canada, Finland, Norway, and Sweden, in what became known as the "Toronto Group" of countries advocating international regulations to reduce CFC emissions. American CFC industry organizations, for their part, objected strenuously to the United States' new position, even though the United States already had an aerosol ban in effect, apparently because an international ban on any CFCs would in effect imply that CFCs were a problem, which would put CFCs in other, more essential uses, at risk of regulation as well (Parson 2003, 117).

Negotiations for a Convention

Just four years after the United States joined the Toronto Group, both the Vienna Convention as well as the Montreal Protocol, which for the first time set concrete international controls limiting CFC consumption, had been negotiated and adopted. Even so, negotiations on the Montreal Protocol resulted in a Protocol that looked nothing like the domestic legislation in effect in the United States when the call for international regulation was first made. Even more remarkable, although the Toronto Group was at first united in calling for a ban on nonessential aerosol uses of CFCs, it was the United States itself that broke from the other Toronto countries in 1986 to bring out a radical proposal for a 95 percent reduction in consumption of *all* CFCs. This shift in US position was, however, compatible with the evolution of US interests during that time. What changed in the interim?

It would be easy to speculate that other countries must have influenced the United States' position during negotiations from 1983 onwards. The Europeans, for example, met the Toronto Group's proposal for international regulation of CFCs in nonessential aerosols with strong opposition. But the European Community (EC) rejected the notion of *any* international ozone regulation, as did the other major CFC-producing countries such as Japan and the USSR and most if not all of the participating developing countries (Benedick 1998, 42[15]; Tolba and Rummel-Bulska 1998, 58). An asymmetric deadlock thus existed (Grundmann 1998, 198; Parson 2003[16]).

The position of the EC shifted in 1984 when it decided on a different tactic to respond to the Toronto Group's proposal. In 1980, the EC had capped production capacity for CFCs at the 1980 level and enacted a 30 percent cutback in CFC aerosol use from 1976 levels (Jachtenfuchs 1990). It therefore countered the Toronto Group with a proposal for an international production capacity cap and a 30 percent cut in nonessential aerosol use of CFCs. There were good reasons for doing this from the EC's point of view. First, the EC definition of "production capacity" in its 1980 decision actually allowed output to increase by over 60 percent from then current levels. Second, like the Toronto Group's proposal, the EC proposal embodied policies already adopted in their own group of countries and was therefore relatively costless for European industry. A 30 percent reduction in aerosol use was trivial in that it could be achieved without much difficulty.[17] Indeed, sales of CFCs for aerosols had already declined in Europe by over 28 percent from their 1979 peak, mainly thanks to unilateral action by West Germany (Bellany 1997, 149). Finally, the EC objected to being forced to adopt regulations that would favor the Toronto Group countries (Morrisette 1989) but felt increasing pressure to do something rather than simply be seen as intransigent (Benedick 1998, 43).

This shift by the EC meant that the EC no longer opposed any agreement, at least on record, but the structure of the situation shifted from one of asymmetric deadlock to one of value-claiming deadlock (Benedick, 43). Value-claiming—or attempting to make the agreement more favorable to one's own self-interest—is common practice in negotiations but can have the effect of eliminating the possibility of reaching any agreement. The question of whose favored measures would be agreed, if any, continued. Whether a case of asymmetric deadlock or value-claiming deadlock, neither situation was conducive to effective agreement.

Some analysts have looked favorably upon the EC's position. Haigh views a limit on total production as the only action that would actually be effective, calling the EC's counterproposal "an intellectually defensible approach which ultimately became incorporated into the Montreal Protocol" (1992). The Toronto Group's proposal would do nothing to prevent releases from growing non-aerosol uses, even though it would be the quickest way of obtaining an immediate reduction in CFC releases from aerosol uses. Bellany (1997) notes that the European ban on CFCs for other uses was taken in advance of the emergence of obvious substitutes, unlike the ban on CFCs in nonessential aerosols. The EC pointed out in its counterproposal that implementation of environmental legislation limiting production rather than use of CFCs would also be much easier to verify since there are comparatively few production centers (Lammers 1988).

Another rationale for the difference in European response has to do with an actual assessment of costs and benefits of each type of international regulation. For Europeans, the benefits of banning aerosols did not outweigh the costs. The US government's acceptance of the science alerting the world to CFCs' role in ozone depletion seemed to the Europeans to be linked to the fact that it came from American scientists. Europeans, particularly the British and the French, were skeptical of US scientific claims, particularly because the theory of ozone depletion was put forth by American chemists who had little background in atmospheric science (Maxwell and Weiner 1993, 21). European government-sponsored research was stressing the uncertainties of ozone science and challenging some of the conclusions of US-based studies.

Meanwhile, the cost of banning CFC use in nonessential aerosols was perceived to be higher in Europe than in the United States. First, the aerosol industry in Europe was larger relative to other users of CFCs, utilizing around three-fourths of all CFCs used in Western Europe (in the United States the figure was half), so a greater proportion of the CFC industry in Europe would be affected than in the United States. The United Kingdom's position was even more exposed, with 80 percent of U.K. consumption of CFCs being used as aerosol propellants in 1974. Per capita, however, the United Kingdom's 1977 consumption of CFCs was roughly half that of the United States, suggesting to many that there was much less need to regulate consumption there than in the United States (Maxwell and Weiner 1993, 21).

Second, the export trade in CFCs was much more important to the Europeans than to the Americans; a ban would entail a loss to them of substantial export markets and a loss to the European balance of payments. This would be exacerbated by the comparative advantage that the Americans had gained in developing substitutes, leading to European fears of having to import increased levels of American-produced substitutes. This assessment of costs was particularly influential in countries such as Britain, where, according to Maxwell and Weiner, "it [is] customary and regarded as legitimate . . . to weigh such economic considerations in decisions on regulation," in contrast to the position of US regulatory agencies that these authors claim are "often unwilling to be constrained by the economic impacts on industry" (1993, 22).

Finally, the British and French in particular envisaged further potential costs from regulation (Rowlands 1995, 106), in the form of possible sabotage of their joint Concorde supersonic transport project through the ozone layer debate. In the early 1970s, scientific studies in the United States had made a tentative link between supersonic transport and ozone depletion. Britain and France had been hurt by this, as the United States not only halted American development of a commercial supersonic transport aircraft (SST) partly as a result of those studies, but also decided to limit access at US airports for the joint British-French-designed Concorde, a move that ultimately destroyed its commercial viability.

While Dotto and Schiff point out that the main environmental issue surrounding the fight over Concorde landing rights in the United States was noise, not CFCs (1978, 92), the American scientific claims were seen by some British and French actors as "a little too convenient for US aircraft manufacturers, after Congress had denied them funds to develop a supersonic transport aircraft of their own" (Bellany 1997, 147). The British and French therefore questioned US motives and were less inclined to regulate CFCs without more conclusive scientific evidence linking CFCs with ozone depletion (Morrisette 1989, 802–806). It was those countries that encouraged their own scientific inquiry into ozone chemistry that effectively produced a different assessment of costs and benefits.

Whether defensible or not, the question is whether Europe's position ultimately moved that of the Toronto Group. The evidence is that it did not. During the last negotiating session for the Vienna Convention the position of the Toronto Group moved slightly, to a proposal for a "multi-option approach" that incorporated four options for action from which parties could choose. Options I and II concerned reduction, in differing stages, of total annual use and exports of CFCs in aerosols, as per the Toronto countries' preference. Option III entailed a freeze on total production capacity for CFCs as from the moment of entry into force of the proposed protocol. Option IV called for a gradual reduction of total annual use of CFCs by 20 percent. None of these matched the EC's proposal in controlling production, use, and exports of CFCs in the non-aerosol sector.

The Europeans were quick to counter the multi-option approach with their own proposal that would set a definitive limit to the production of at least certain CFCs, and the gap between the two groups could not be bridged at Vienna (Lammers 1988, 227–230). In March 1985, the Vienna Convention to Protect the Ozone Layer was signed by 20 countries, the United States and the EC countries among them. It created a general obligation for nations to take "appropriate measures" to protect the ozone layer, a phrase left undefined. The goal of a simultaneously negotiated annex laying down specific limits on emissions of ozone-depleting gases, first proposed by the Toronto Group in 1983, was not achieved (Tolba and

Rummel-Bulska 1998, 60). However, in Vienna a resolution was tabled by the United States to authorize UNEP to reopen negotiations with a 1987 target for arriving at a legally binding control protocol, which passed over the objections of most EC countries (Benedick 1998, 45). While the European counterproposal is a case of value-claiming, according to the framework used here, in fact both sides were equally guilty of value-claiming, given that their respective positions were heavily influenced by the positions of their respective industries and their competing desires to "internationalize" the respective policies each had already taken domestically (regionally in the case of the EC) so as to cause least disruption to themselves (Lammers 1988, 230; Rowlands 1995, 111–112; Downie 1996, 303). Although there was no longer asymmetric deadlock among CFC producers, because all major parties were on record as advocating international controls, there was certainly a stalemate as to how to proceed. Such was the case until other factors intervened in 1986 to shift the United States' position.

Stage Two: The Outcome in Montreal

With the stalemate over how to regulate CFCs, negotiations were renewed in December 1986, as had been called for in the resolution taken in Vienna. These negotiations resulted in the signing of the Montreal Protocol, the first major breakthrough in controlling ozone-depleting substances, in September 1987. The Protocol was completed with remarkable speed, entering into force on schedule, on January 1, 1989, just three months after the Vienna Convention itself entered into force.

From a 1986 baseline, the Montreal Protocol called for a 50 percent reduction in calculated production and calculated consumption of CFCs by 1998, along with a freeze on halons, another ODS, by 1992, for developed countries. For developing countries those deadlines were postponed by a ten-year grace period, thereby allowing production in those countries to keep growing for ten years to meet "basic needs". While not perfect, the Protocol successfully set targets and timetables for controlling environmentally destructive human activities and became a model against which to compare all succeeding international environmental agreements.

Many analysts attribute the relative speed and ease of passing the Montreal Protocol to an increasing scientific consensus on the causes of ozone depletion (see, e.g., Haas 1992a; Tolba and Rummel-Bulska 1998). Haas bases his argument regarding the positive influence of epistemic communities on the assumption of the preeminence of science in influencing positions during negotiations for the Montreal Protocol. Most significantly, the discovery of a "hole" in the ozone layer over Antarctica in 1984, by a team of British scientists (Farman et al. 1985), led to an increase in concern over the extent of damage being done.

On the other side, Weiss goes so far as to state that "the role of science in this controversy was extremely limited" (1993, 232). This is backed up by Benedick's recollection that although rumors of the discovery circulated for months before its publication, it had no effect on the negotiations for the Protocol. According to him, Antarctica was never discussed at the negotiations, and even two months after Montreal the EPA concluded that it could not yet serve as a guide for policy decisions (1998, 20). Indeed, policy makers in Britain, a key veto state, regarded the Antarctic ozone hole as "an anomaly that underscored the inadequacies of existing scientific theories" (Maxwell and Weiner 1993, 29). The hole was not stressed before or during Protocol negotiations because there were so many competing explanations for the hole that the negotiators could not assume it was related to their task (Rowlands 1995, 56; Benedick 1998, 20).

Shimberg points out that all the negotiators were at least aware of the harmful developments over Antarctica, so it may be assumed that each made some subjective assessment of the scientific evidence, but the fact that there was still no established scientific knowledge at the time of the signing of the Protocol casts doubt on any argument that science alone can explain the success in achieving agreement on the Montreal Protocol (Grundmann 1998, 206–207; Parson 2003, 142). No available scientific theory could explain the loss of ozone over Antarctica. Although three competing hypotheses were soon advanced, only one of these cited a possible relationship between CFCs and the ozone hole.[17] As late as 1986, the ARCFCP was able to argue that "some reduction would be occurring in the ozone layer over the Antarctic even if CFCs were never introduced into the atmosphere" (Alliance for Responsible CFC Policy 1987, iv). The theory of ozone layer depletion, and in particular the seasonal decrease in ozone over Antarctica, was not confirmed until 1988, with a report from the Ozone Trends Panel, a 100-member research group formed by NASA in 1986 to reevaluate existing long-term ozone data. Nevertheless, even though it is unclear that science played a direct role in the negotiations, the new scientific knowledge and any sense of crisis that grew out of it can be seen as one of several factors that caused a general shift in the economics of ozone for the United States during the mid-1980s.

Shifts in US Interests

The US negotiating position on international CFC regulation shifted in 1986, from advocating a ban on nonessential aerosol uses of CFCs to calling for a long-term scheduled reduction of up to 95 percent of all CFC and halon consumption. This evolution in the US position cannot be fully explained by science, nor is there any claim that the European proposal for capping CFC production had any influence on the United States. The new position correlated closely, however, with a shift in the economics of CFCs for the United States. Indeed, US government documents and memoirs indicate that a weighing of costs and benefits was paramount for many of the officials involved.

First, a draft EPA study came out in 1986,[18] which was the most detailed and influential assessment of costs and benefits of policy action and inaction on ozone depletion undertaken up to that point. There had been "substantial learning" on the effects of failing to manage emissions of ozone-depleting substances since earlier CBAs (Dudek and Oppenheimer 1986, 357). Despite difficulties in comparing long-term benefits with short-term costs, the EPA's estimates of the costs of inaction were significantly higher than earlier estimates, including increased incidence of skin cancer, cataracts, blindness, as well as immune system suppression, damage to agriculture and fisheries, and other effects. In fact, numerous CBAs of ozone regulation were carried out for the United States in the mid-1980s, many of which studied the ozone depletion problem in terms of the costs and benefits of avoiding skin cancer resulting from ozone depletion and increased UV exposure. Even limiting the analysis of potential benefits of action to this easily quantified benefit and using an 11 percent discount rate for calculating future costs and benefits, the benefits of action were still found to outweigh the costs.[19]

During congressional debate on the US negotiating position with regard to the prospective protocol in the spring of 1987, Senator Steve Symms (R-Idaho) weighed these benefits against the spectre of lost American jobs, lowered standard of living, and reduced international competitiveness from CFC regulation, as well as the questions still surrounding ozone science and the availability and cost of safe substitutes.[20] In their testimony, however, representatives of the EPA were able to draw upon the EPA study as well as numerous others "that showed the overwhelming positive cost-benefit ratio associated with elimination of CFCs and related compounds."[21] This resulted in a Senate vote approving the original US negotiating position calling for a 95 percent reduction by 80 to 2 on June 5, 1987.

The President's Council of Economic Advisers also issued a cost-benefit study in 1987, which showed that the monetary benefits to the United States of preventing future deaths from skin cancer alone "far outweighed the costs of CFC controls as estimated either by industry or by EPA" (Benedick 1998, 63). According to Benedick, this conclusion was reached based on the most conservative estimates, without attempting to quantify other potential benefits of preventing ozone layer depletion. The chief US negotiator himself called this study "a major break in interagency debate" over the US negotiating position (63).

But it was the American CFC industry that mattered, and gradually its position began to shift. Until 1986, the industry argued that CFC markets were already stagnant due to the United States' unilateral regulation of CFCs in nonessential aerosol uses, so there was no need for further regulation (even though in reality CFC markets and production began to grow again after 1983). However, in September 1986 the ARCFCP issued a public statement declaring support for a "reasonable global limit" on CFC growth.[22] The Alliance statement cited mounting scientific evidence of stratospheric ozone depletion as the reason for the change in policy. This was the first industry acknowledgment of the possible connection between CFCs and stratospheric ozone depletion. Until the mid-1980s, the American industry's position had been that the scientific evidence connecting chemical manufacture to ozone depletion was too uncertain to warrant international regulation. However, in 1986 NASA and the WMO published a scientific assessment indicating worsening atmospheric conditions. The NASA/WMO assessment was cited as the rationale for the industry shift.

Some analysts take the industry claim that science was behind this shift in position at face value.[23] However, Benedick notes that during the Protocol negotiations some Europeans, specifically, "suspected that the US was again using an ozone scare to cloak commercial motives" (1998, 33), and the assertion that the conclusions of this study were more ominous than those of past studies has been questioned (see Litfin 1994, 93; Parson 2003, 157). Rowlands (1995, 113) points out that it is impossible to identify any effects of the new scientific knowledge in isolation from the other considerations that more directly faced industry at that time.

Moreover, evidence shows that the industry's new position favoring international limits was due not only to a desire to address the greater environmental costs of ozone depletion that were now being predicted but also to the perception that other potential benefits of international regulation were also increasing, while its costs were decreasing.

One major benefit to industry would be to avoid the threat of congressional legislation that would unilaterally restrict non-aerosol CFC production and use. In October 1986, Senator John H. Chafee (R-Rhode Island) called for consideration of further legislation to control CFCs and other ozone-destroying chemicals in light of the international negotiations scheduled to begin that December:

> [I]f it appears that this round of negotiations is not likely to produce a strong protocol phasing out fully halogenated CFCs . . . we would have to reassert our role as national decision makers and move legislation forward to require unilateral action . . . leading to the phase-out of such CFCs.[24]

Sebenius sees this threat as an example of action to achieve a change in the perceived zone of possible agreement for American CFC producers (1992b, 357), because although no regulation was industry's ideal alternative, international rules that would constrain the entire global industry were far preferable to a US law that would only apply to domestic companies in the absence of international regulations.

Domestic legislation on non-aerosol CFC production and use threatened a cost to the industry that the earlier legislation on nonessential aerosol use had not done, because of a difference between the markets for US-produced CFCs for nonessential aerosol uses and those for other uses. The market for nonessential aerosols was almost entirely domestic (Benedick 1998, 27), and because the ban on these regulated use rather than production, it covered imports into the United States as well. This meant CFC substitutes in this area were not vulnerable to imports of CFCs from non-regulated countries, and the manufacturers that had had to switch to substitutes at that early stage were not at an international competitive disadvantage. On the other hand, CFCs in "essential" uses went into products that were much more vulnerable to international competition. Firms in the United States were alone among the five CFC producer countries' firms in facing stringent domestic CFC regulations, as well as the threat of more regulation, prior to negotiations on the Protocol itself (Baldwin 1999, 113). This threat of domestic legislation on CFC production for essential uses thus gave manufacturers an incentive to support some form of international regulation over all CFCs.

The industry also had an interest in avoiding unilateral regulations being forced upon the EPA. The EPA had been mandated by the 1977 amendments to the Clean Air Act (Doniger 1988, 88; Litfin 1994, 72; Rowlands 1995, 114) to regulate non-aerosol CFC uses but had never done so, due to the flood of objections from industry. Litfin (1994, 93) notes that these objections had been predicated on the fact that CFC use in the United States had already dropped precipitously in the 1970s with the ban on nonessential aerosols and on industry's own projections of no future growth in their use. However, these projections were proven untenable by statistics showing CFC use rising back to levels seen prior to the 1978 ban as the global recession of the early 1980s ended. By the mid-1980s the downward turn in sales of CFCs that had begun in the mid-1970s had been reversed, and global CFC production, excluding China, the USSR and Eastern Europe, exceeded 1974 production (Shea 1988, 24). In the United States in particular, roughly one-third of the world total of CFCs was being produced and per capita usage of CFCs was among the highest in the world despite the US ban on nonessential aerosol uses (Shimberg 1991, 2184); hence the push for the EPA to finally fulfill its obligation. As a result of a lawsuit brought by an American

NGO, the Natural Resources Defense Council (NRDC), the EPA agreed in 1985 to start formulating regulations on non-aerosol CFC uses, as per its mandate (Doniger 1988, 88; Litfin 1994, 72; Rowlands 1995, 114).

Lawsuits against CFC manufacturers themselves could also be avoided with international regulation. Lawsuits were beginning to be brought against cigarette manufacturers, so this potential threat seemed real. In addition, by supporting international regulation, companies could gain a benefit for their corporate image from presenting themselves as good global citizens, thereby avoiding more consumer boycotts (Rowlands 1995, 113; Grundmann 1998, 213).

But an agreement to regulate at the international level would provide even more tangible economic benefits for industry as well, in opening the possibility for expansion of US export markets, a point which Senator Chafee himself highlighted in his call for legislation. Because of the early incentive it would provide for domestic industries to begin work on producing safe alternatives to the production and use of ozone-destroying compounds, American industry "would get the jump on its competition abroad,"[25] and would then be able to export CFC substitutes to assist others in achieving CFC reductions.

As early as 1979 it was known that research and development into new substances had begun in the United States in anticipation of further regulations, and a number of potential substitutes for CFCs had been identified (Rowlands 1995, 114). Testing began in the early 1980s as part of the normal seven-to-ten-year process of bringing new products onto the market. Progress was then halted as Reagan's ascension to the presidency made the likelihood of regulation recede, taking with it the prospects for markets for the new chemicals. However, with new scientific evidence research on substitute chemicals was started up again, and despite intraindustry secrecy there were numerous clues that progress was being made even before the Protocol was signed. In particular, industry's claim that viable CFC alternatives did not exist for most uses was undercut at a scientific workshop sponsored by the EPA in Leesburg, Virginia, in March 1986, where details came out showing progress on researching alternatives for most CFC uses (Parson 2003, 125–126).

On the cost side of industry's balance sheet, the fact that the United States had already eliminated aerosol uses meant that it was more difficult for US industry to reduce CFC production than it was for its competitors in Europe (Litfin 1995). However, the perceived costs of regulation were falling because of progress in research on substitute chemicals for non-aerosol CFC uses, from whose marketing some of the same industrial actors stood to benefit. The US industry association's initial estimate of costs for implementing the Montreal Protocol as it was finally agreed in 1987 was between \$5 and \$10 billion (Alliance for Responsible CFC Policy 1987, I-8). As little as one year later, though, it was "generally accepted that a substantial total reduction—at least

50 percent—in combined CFC and halon use could be accomplished relatively quickly and at little cost" (Benedick 1998, 119). By 1990 the *Economic Report of the President* estimated US costs of compliance with the Protocol at $2.7 billion; later this was reduced even further (Sebenius 1995, 53).

Industry's change in position was thus motivated not just by changes in scientific knowledge but by the impacts of these and other changes on the economics of the situation for the companies involved. It is safe to say that the fact that American industry came on board by the late 1980s was the catalyst for the effective outcome of the negotiations for the Protocol, even if not the driving force behind it. Although the US government had called for international regulation of CFCs in 1977 and specifically for a ban on CFCs in nonessential aerosols in 1983, it was not until the 1986 Alliance declaration that the US industry gave public support to international regulation of CFCs. It was only after this that the United States' position shifted and broke the deadlock. The idea that the international of CFCs was pushed by American industry (DeSombre 2000, 27–28) is therefore not supported, but there is no question that the Montreal Protocol was only successfully concluded after US industry's "change of heart."

But there was one more factor at work. Just ten days after the industry association's announcement, DuPont broke ranks with the rest of the industry and called for a limit on the *level* (as opposed to growth) of worldwide CFC production at the UNEP negotiations (Roan 1989, 192). This caused a split in the industry and confusion among those who believed that CFCs were not posing a major problem (Grundmann 1998, 212), given DuPont's status as one of the largest and most influential actors in the CFC industry. DuPont garnered a full 25 percent of the global market in CFCs as of 1988 (Thomas 1992; Smith 1998) and over 50 percent of the US market, a consequence of its role in the discovery of CFCs in the 1920s—indeed, DuPont held the patents for CFCs until they expired in the late 1940s (Metzger 2005).

With this action DuPont in effect sacrificed the other manufacturers by splitting industry's united front and thus significantly weakening industry's position of influence (Sebenius, 1992b, 358–359; Parson 2003, 127). Once again, the 1986 NASA/WMO assessment was cited.[26] However, there were probably other reasons behind DuPont's shift. The fact that DuPont maintained proprietary rights to its CFC substitutes suggests that a desire to address the scientific threat did not outweigh the profit motive for DuPont (Smith 1998). Parson, however, notes that DuPont's $600 million CFC business accounted for only 2 percent of its sales and was therefore not particularly significant to it. While smaller CFC producers were more threatened by CFC controls, the larger industry actors, including DuPont, may well have started to realize by this point that regulations to restrict CFCs might bring positive results for them in the form of market consolidation and more favorable market conditions (Parson 2003, 127).

In addition, DuPont appears to have been in an advantageous position even relative to the two other largest American CFC producers, thanks to its work in developing CFC substitutes. A DuPont representative indeed said as much in a private communication to a customer that September,[27] and it was generally thought that DuPont was also well ahead of its global competitors in the search for substitutes.[28] International limits on production and consumption of CFCs would thus place DuPont in a very favorable competitive position. While DuPont had issued a paper in February 1986 that concluded that "fully satisfactory fluorocarbon alternatives will not become available in the foreseeable future" and stated that "no further studies . . . are planned by DuPont" (cited in Miller 1989, 549), it was DuPont itself that at the March 1986 Leesburg workshop announced what amounted to a concession that substitute chemicals for CFCs were available.[29] Summarizing its work on substitutes, DuPont stated that it could develop substitutes in about five years but they would cost between two and five times as much as CFCs and that "neither the marketplace nor regulatory policy . . . has provided the needed incentives" to justify the required investment.[30] In other words, the unavailability of substitutes was a function of the absence of a market and price incentives rather than of the chemistry required.[31] According to two DuPont officials writing later that year, what was needed was pressure to develop alternatives.[32] International limits would permit the price of the allowed production of CFCs to be raised, the implication being that this price increase would be welcomed by CFC manufacturers that were not dependent on producing a minimum amount of the substance to maintain profitability (Sebenius 1992b, 358).

Analysts agree that DuPont was a major player in the controversy over CFCs and the ozone layer. DuPont greatly influenced the US position in the international arena when it shifted from being "the most visible defender of CFCs" to become a proponent of stricter controls than the rest of industry was willing to accept (Grundmann 1998, 212). Furthermore, because of its huge size and market share, it might be argued that what was good for DuPont really was good for the United States, to borrow a phrase. DuPont's position as the leader in CFCs put the United States in a position to break ranks with the other Toronto Group countries and push for regulation of all CFCs. Sprinz and Vaahtoranta (1994) note that the first reports about the development of new substitutes for CFCs appeared in the press at about the same time that the US government began to strive for ending all uses of CFCs and that it was generally believed that the new position of the United States was bolstered by success in developing new forms of chemical compounds.[33] Alan Miller asserts a different direction of causality, saying that it was the likelihood of regulation that made DuPont and others look into development of substitutes in the mid-1980s after a hiatus in research (1989, 548), but this leaves aside the fact that it was only after DuPont's announcement about substitutes in March that the United States changed its position on regulation of CFCs in "essential" uses.

After the first DuPont announcement in March of 1986, US negotiators began to "sell" the idea of a full 95 percent cut in CFC consumption (Benedick 1998) and then tabled a proposal calling for an ultimate 95 percent reduction in emissions of a selection of CFCs and halons at the first round of negotiations for the Protocol in December 1986 (Lammers 1988, 232). Parson notes that there was some speculation that DuPont had actually contributed to this extreme position through the secret development of substitutes; he refutes this claim, however, arguing that, among other things, it would have been impossible to develop commercial synthesis processes, prove their suitability for certain applications, and test their toxicity completely in secret (Parson 2003, 126).

But if DuPont was not behind the United States' "extreme" 95 percent reduction proposal, what was? According to Parson, "a group of US officials, primarily within the EPA, who were strongly concerned about ozone" were once again responsible for the US position, as had been the case during the United States' domestic "ozone war" of the 1970s (2003, 122). During the course of the 1987 negotiations, the United States came to accept a 50 percent reduction in production and consumption by 1998. There are factual discrepancies in the various reports of how this US position evolved. Benedick downplays the discrepancy between the original proposal for a 95 percent cutback and the United States' ultimate acceptance of only a 50 percent reduction in the Montreal Protocol. He asserts that the initial figure of a 95 percent cut was "illustrative only," because full agreement had not yet been reached within the US government on the extent and timing of CFC reductions. He notes that Congress and others outside the executive branch "often chose, however, to interpret the 95 percent target as implying a firm commitment to virtual phase-out" (1998, 54). According to Benedick, the United States promoted bracketed language calling for a reduction in CFCs of a range of 10–50 percent in an unspecified number of years during the second negotiating session (1998, 58). Benedick thus portrays the US position on CFC reductions as flexible from the first negotiating session for the Protocol.

Other analysts, however, portray the US position as shifting as a result of industry pressure between the second and third negotiating sessions. Lammers (1988), in his very detailed summary as an observer of the first and second rounds of negotiations for the Montreal Protocol, makes no mention of any change in the United States' negotiating position on this point. Indeed, Doniger an American NGO observer at the negotiations for the Montreal Protocol, states that the United States elaborated upon its original proposal for 95 percent cuts at the second negotiating session in February 1987 by suggesting a 10- to 14-year period for the reduction, but then "put little, if any, emphasis on securing reductions beyond 50 percent" at the third negotiating round (Doniger 1988, 89). Newspaper reports of that era support Doniger's figures in general,[34] speculating that the US position was softening during the third round.[35] Litfin also suggests it was at the third meeting that the US position had changed, quoting US official David Gibbons as saying that at the third negotiating session in April 1987, the United States, "was decidedly less vocal than at previous meetings" (1994, 113).

Litfin notes that no segment of US industry ever supported the original US position calling for a 95 percent cut, while Doniger states that the US chemicals industry lobbied for a lower reduction before the third round of negotiations in April 1987 and received help from anti-regulatory elements in the White House and the Departments of Energy, Interior, and Commerce. This, according to Doniger, was the point at which the American delegation changed its position.[36] Parson confirms that it was indeed CFC producers and consumers in the United States that caused the United States to back down before the third session, helped by the fact that only the United States was pushing for anything more than a freeze in CFC use in the international negotiations (Parson 2003, 134). Industry's strategy was to organize and gather support from officials in the US Administration and Congress who would ally themselves with industry's position and rein in the officials who were pushing for such a huge reduction.

A bitter contest ensued among the various sides within the US government (2003, 133–135). Senator Chafee introduced legislation in February 1987 that would require a 95 percent reduction in new production within six years, and then in June 1987 a Senate Resolution was passed urging the United States to return to its *original* negotiating position calling for virtual elimination of certain CFCs (a fact Benedick fails to mention). The Resolution was nonbinding and did not affect the final outcome in Montreal three months later (Shimberg 1991, 2188), but after the Protocol was signed the US legislation was modified to reflect exactly what had been agreed in Montreal (Doniger 1988, 91).[37]

In the end, the Montreal Protocol of 1987 reflected the 50 percent compromise figure mentioned above was the 50 percent figure simply the result of "splitting the difference" between the hard-liners demanding 95 percent cuts and the states and other actors that argued for a freeze only, as Parson argues (Parson 2003, 144)? Or was always a "bottom-line" position for the US negotiators, who, according to Benedick, had only used the extreme position as an illustration and who, according to Parson, managed to keep others guessing about their commitment to it until late in the negotiation (2003, 143)? In either case, it seems that the United States' acquiescence in the 50 percent compromise was influenced by US industry.

The shifts in the US position on international ozone regulation are attributable not solely to the negotiating process or to the science of ozone depletion but to a combination of factors that led to shifts in the relative values to the United States of an agreement relative to a no-agreement outcome. Although the science of ozone depletion was influential at least to some extent in shifting the positions of both industry and government regarding CFC regulation, an increasing perception of threat from ozone depletion was aided by other industrial considerations. The unfolding of events in the United States is consistent with the view that it was shifts in perceptions of the relative value of benefits versus costs of agreement for the United States that ultimately led to an agreement for a 50 percent reduction in CFCs in the Protocol as adopted in Montreal. The greater the scientific threat, the greater the chance that the benefits from avoiding it will outweigh the costs, but this is not guaranteed.

The United States versus Europe, 1987

With shifts in benefits and costs now favoring international regulation of at least a 50 percent cut in CFCs, the United States was now willing to use both sticks and carrots to win anti-agreement actors over to agreement. The first level of dispute was mainly between the EC and the United States. Jachtenfuchs, for example, calls the negotiations that began in December 1986 "quasi-bilateral negotiations between the EC and the USA" (1990, 264).

Some Europeans highlight what they see as European leadership in these talks. Haigh, for example, points out that the EC was the original architect of the production limit, and therefore "won the intellectual argument."[38] Elsewhere Haigh (1992, 246) also points out that if the EC had not prevented the Toronto Group's original proposal for a ban on CFC use in nonessential aerosols from being adopted in 1985, the protocol would have been less satisfactory, needing complete revision after the discovery of the Antarctic ozone hole and probably lacking full support from all important CFC producer states. This is supported by the statement of an EPA staffer that with an aerosol protocol it might have taken another ten years to "fill in the other ten big uses."[39] Parson, on the other hand, argues that the ultimate

50 percent reduction cut required by the Montreal Protocol was the result of bargaining rather than science, a compromise between the EC's proposed freeze and the United States' 95 percent cuts.

Neither of these views is entirely convincing, however. First, the US negotiating position of a 95 percent reduction in CFC consumption, beginning in December 1986, was not favored by the Europeans and was of course not a result of pressure from the Europeans. The eventual agreement for a 50 percent cut in consumption of CFCs looked like the United States' position after pressure from the American CFC industry, not like the European proposal for a cap on production capacity, and it was the United States that exerted pressure on the EC to accept this compromise, not vice versa. Moreover, the Europeans found it hard to swallow: Jachtenfuchs reports that the common EC negotiating position broke down during the second round of negotiations over how far to go beyond a 20 percent reduction (1990, 266).

As reported in British newspapers of the time, Britain, with its lucrative CFC industry led by Imperial Chemical Industries (ICI), dictated a slow pace in negotiations and opposed the EC making any concessions until the third round of negotiations in May 1987.[40] Numerous observers and commentators attribute the backdown first and foremost to pressure put on Britain by the United States. This pressure took its most concrete form in the proposed US legislation introduced by Senator Chafee in February 1987. Not only did Senator Chafee's proposal call for unilateral CFC cuts of 95 percent by the United States by January 1, 1993, it also mandated unilateral trade sanctions against products containing CFCs "to assist our negotiating team" and "to prevent other countries from taking advantage of unilateral United States action."[41] These sanctions were intended to lessen the international competitiveness-reducing effects of unilateral regulation of CFCs in the United States, and particularly threatened harm to other states that produced goods containing CFCs for the US market.

In the end, Congress did not actually vote on the proposal to restrict entry of goods made with ozone-depleting substances once it became clear that multilateral efforts to regulate them were moving forward.[42] However, the introduction of such legislation was clearly intended to shift the cost of no agreement upward for opposing countries and thereby influence their preferences regarding an effective agreement. The threat it posed became real during the negotiations when Benedick, as lead US negotiator, criticized EC proposals accepting only 20 percent reductions over a long term as "ridiculous" and "totally unacceptable" and, in the closing press conference, stated that the US Senate was "getting impatient. If the United States takes a unilateral decision to ban the CFCs, it will be accompanied by a trade embargo of these products, such as European ones, which include these products."[43]

Was it US pressure that led to the compromise? According to the US Chamber of Commerce it was a key factor in advancing international negotiations.[44] Both the credibility of the threat and the outcome in Montreal indicate so. First, there was no reason to think that the threat was not credible. The proposal was made some seven years before the establishment of the World Trade Organization would throw into question the legitimacy of such unilateral threats, and as it only included products containing CFCs rather than goods produced from processes using CFCs, the threat of sanctions could be taken as credible by the parties most likely to suffer from them. In addition, the idea that the United States was prepared to take unilateral measures against CFCs, including trade sanctions, was bolstered by the fact that in its proposal for international regulation the United States was calling for measures for which it would bear more costs than would other countries, which it was clearly prepared to do (Parson 2003, 143).

As for the results of this threat, Parson reports that "the serious threat to trade in CFC-related products clearly weakened the resistance of the Japanese and Europeans" (2003, 143). On the other hand, he notes that Benedick's remarks antagonized many delegates at the time; indeed, one European delegate rejected the United States' criticism, noting the fact that CFC production had actually been continuously increasing in the United States despite its ban on CFCs in aerosols.[45] However, Benedick, in his own account of the negotiations, says observers later commented that his statements "contributed importantly to building public pressure in Europe for a stronger treaty" because of the wide media coverage they received (1998, 71).

Benedick associates himself with the "personal text" produced by Tolba during small, closed meetings during the third negotiating round, saying it "increasingly resembled the original US proposal" (1998, 72), despite the fact that it began to converge around the 50 percent compromise figure on CFC cuts. Benedick also notes a look of "stunned disbelief" on the faces of many European delegates as the Protocol was adopted in Montreal (1998, 76). This, along with the conclusions of other analysts that it was "strong US pressure" and "political muscle" that brought the EC to agreement on a 50 percent reduction on consumption and production of CFCs (Jachtenfuchs 1990, 267; Haas 1992a, 207) leave little doubt that the 50 percent figure was something the United States was satisfied with.

Morrisette attributes eventual European acquiescence in a 50 percent reduction in CFCs between the second and third rounds of negotiations in 1987 not only to "difficult negotiations" and the fear that the United States might take unilateral action and impose trade sanctions, but also in part to the weight of recent scientific evidence and pressure from European environmental groups (1989, 811). The United States' efforts were certainly helped by pressure the West Germans and Danes began to put on Britain, as the "dirty kid" of Europe, in mid-1987,[46] although it might be questioned how effective the Germans would have been in achieving an effective treaty in the absence of US support, given later experiences on climate change.

It also helped that there were other incentives for the Europeans. The new American position of 1986, favoring across-the-board cuts rather than cuts in aerosols only, actually provided a double inducement to the EC. In addition to the threat of punishment through trade sanctions that would effectively lower the value of no agreement for the EC, the new proposal also raised the value of agreement for the EC. A uniform percentage reduction in CFCs favored the Europeans because they could cut back in the most "cost-painless" sector of aerosol usage, while the Americans would have to achieve reductions in other applications, having already eliminated CFCs in aerosols. In addition, because European usage of CFC-propelled aerosols was very high (higher than even in the pre-ban United States), all of their reductions could be made from cutting aerosol usage. Thus the agreement favored the Europeans relative to the Americans; this represented an enticing "carrot" (Rowlands 1995, 115) that raised the value of the agreement for them.

Moreover, European chemicals industry giants also faced the promise of expanding profits from marketing substitutes. The new substances required very large capital investments, but they could be marketed as high-margin specialty chemicals, would thus command higher prices, and ensure a substantial competitive advantage to a few leading international firms that could make that investment. This would particularly benefit ICI, the largest British CFC manufacturer as well as one of the largest European manufacturers. But this could not happen without government intervention to control CFC use, as the projected price of the new chemicals was five to ten times the cost of the CFCs they would replace. Government involvement could also insure an orderly staged transition to the alternatives (Oye and Maxwell 1994, 600). The fact that Japan, another producer country, expressed concern over the prospect of monopolies on technological information on substitute chemicals at the third round of negotiations is an indication of how important this competitive advantage was (Benedick 1998).

The cumulative result of these shifts was a European revaluation of the benefits and costs of CFC regulation, with the outcome that the Europeans capitulated in a 50 percent reduction of production and consumption of CFCs in time for the September 1987 completion of the Montreal Protocol. The finally agreed text of the Protocol called for a freeze by 1989, an automatic 20 percent cut by 1993, and then, unless opposed by a two-thirds majority of parties representing at least two-thirds of total consumption, a 50 percent reduction by 1998. A compromise was achieved over whether production or consumption would be controlled, with the Americans winning a provision limiting consumption through measuring it as "production plus imports minus exports." This represented a victory for American industry, which wanted to preserve as much of its production of CFCs as possible through reallocating it from domestic consumption to exports[47] (once again calling into question American CFC producers' claims that their actions were based on the science of predicted ozone depletion). Furthermore, production for export (in industrialized countries only) could increase to meet developing country parties' "basic domestic needs" (see below).

The Montreal Protocol was achieved after the value-claiming deadlock between the EC and the Toronto Group—each advocating measures it had already taken domestically—was overcome with an entirely new proposal for a cut in all CFCs, not just those in aerosol uses. What is clear, however, is that without the United States leading the push, including a strong initial position and a willingness to threaten unilateral trade sanctions, the result in Montreal would not have been the effective international agreement that it was. The fact that the United States won, on balance, what it wanted both in terms of stricter controls than its key opponents wanted, as well as in terms of looser provisions in other areas that would favor its own industry, is an indicator not only of the United States' power but also of its configuration of interests on this issue.

Stage Three: Full Phase-out of CFCs

Binding commitments to a full phase-out of CFCs by 2000 were made within three years of completing the Montreal Protocol; by 1992 this deadline was pushed forward to 1996. The explanation for this sudden consensus can be found in continuing scientific and technological developments that shifted cost-benefit calculations for all the developed countries involved. By now, CFC substitutes were already beginning to appear on the market (Rowlands 1995, 117) so the costs of action were decreasing. Meanwhile, the reverse was true for the costs of inaction. In March 1988, the Ozone Trends Panel announced its conclusion that chlorine chemicals were indeed the primary cause of the ozone depletion measured above Antarctica. Then in May a Canadian researcher reported a discovery of ozone losses over the Northern Hemisphere.[48] Early models had predicted ozone losses in the long term; now it was clear that ozone depletion was already occurring.

A further scientific development, this time not a discovery but an improved method of reporting, also provided new information that contributed to the desire for stronger regulation. The new approach used "chlorine-loading potential" (CLP), first defined in 1986 by John Hoffman of the EPA, in order to measure the abundance of ozone-depleting materials in the atmosphere. The CLP was intended to aid comparisons of ozone-depleting substances without making assumptions about their quantitative relationship to ozone depletion, a need that was recognized thanks to the fact that the ODP had been unable to predict the losses of ozone over the Antarctic.

The CLP of an ODS is based simply on the number of years an ODS will remain in the atmosphere, or its "atmospheric lifetime." The CLP approach became attractive after chlorine had been confirmed as the main cause of the Antarctic ozone hole discovered in 1984. While ODP measures the point at which chlorine concentration peaks in the atmosphere, CLP measures levels of chlorine concentrations above the atmosphere's natural level of 0.6 ppb and the length of time it will take to recover from peak ozone loss to more normal conditions. With the CLP, the new benchmark for evaluating future control strategies became the point at which atmospheric chlorine concentrations would return to no higher than 2 ppb, the level at which the Antarctic hole appeared. Projections of emissions that used this new metric indicated that in order to stabilize chlorine in the atmosphere at 1985 levels all CFCs would have to be eliminated, and other compounds containing chlorine and bromine would also have to be brought under some controls (see chapter 4).[49]

An examination of the global economic benefits and costs of regulation was performed in 1989 by the four Assessment Panels set up under the Protocol. On the benefit side, the Panel for Economic Assessment identified "enormous beneficial impacts [at the global level] on human health and the environment" of a full phase-out, the monetary value of which was "undoubtedly much greater than the costs of CFC and halon reductions." The Technical Assessment Panel concluded that it was technically feasible to phase out some CFCs by 2000, although the Panel acknowledged that costs would depend on the rate of technological progress, including "capital costs, research and development costs, safety and toxicity risks, and operational costs" (UNEP 1989b, 12–13). That further commitments depended on technological developments is demonstrated by the fact that policy-makers at MOP-1 in Helsinki agreed to a nonbinding resolution "to phase out the production and consumption of CFCs controlled by the Montreal Protocol . . . not later than the year 2000" but did not extend this commitment to halons or to those CFCs or other ozone-depleting substances for which substitutes were not available (UNEP 1989a).

The Pro Coalition: DuPont and ICI

As in earlier phases, the positions of governments of large CFC producer states primarily reflected the positions of their dominant chemical manufacturers during this period. The evolution of the positions of two of these dominant actors, DuPont of the United States and ICI of Great Britain, in the aftermath of the Montreal Protocol is illustrative. Immediately after the Montreal Protocol was signed in 1987, DuPont claimed that "any substitute will take about seven years to develop." However, in early 1988, DuPont was put under pressure to honor a pledge it had made in the 1970s to halt its manufacture of CFCs if the science demonstrated that they were a credible threat. That scientific evidence appeared with the March 1988 OTP announcement. By the end of that month, DuPont had decided that it was feasible to phase out all production of CFCs, as well as halons, another substance covered by the Montreal Protocol, according to the same long-term timetable for a 95 percent reduction by 2000 that the United States had initially proposed in December 1986. DuPont also announced a shift to alternative chemicals, two of which had already been developed for CFC-12, the "chief offender" in the CFC family.[50]

As in 1986, DuPont again asserted that its decision was driven by the science (Litfin 1995, 267), and Haas insists that this decision entailed real costs and losses in opportunity for DuPont. However, once again, the OTP report was, according to Parson, "less novel and definitive than DuPont's rationalization of its policy change claimed" (Parson 2003, 157), and Litfin notes that DuPont's statement was met with cynicism both by environmentalist groups and by European industry. DuPont had an interest in avoiding the costs of inaction, including damage to company reputation in an era when the importance of cultivating a "green corporate image" was growing,[51] as well as any costs that might be associated with a new lawsuit that the NRDC had filed against the EPA to force them to strengthen regulations.[52]

The industry emphasized the cost transition to new chemicals: around $5.5 billion between 1990 and 2010, rising to $27 billing by 2075. At the national level, however, this argument was not convincing in comparison to the EPA's estimated savings of $6.5 *trillion* in medical costs by 2075, thanks to the Montreal Protocol's expected reduction of skin cancer deaths.[53] The costs of a phase-out were in fact falling as the development of alternatives was progressing rapidly, while the prospect of big profits to be made from international marketing of CFC substitutes was increasing (Rowlands 1995, 118). With the research it had under its belt, DuPont was well positioned to seize the "golden marketing opportunity" awaiting the first company to devise CFC substitutes (Litfin 1995, 267).

Although DuPont was the first to announce a phase-out of CFC production, it was joined by ICI in August 1988, when ICI found itself in a position to take advantage of the new market for substitutes (Baldwin 1999, 115). The British company then also called for an urgent international review to strengthen the provisions of the Protocol toward phasing out CFCs

(Benedick 1998, 118). Until 1987 the CBAs of ICI had been based on British studies that were far more skeptical of ozone science than American ones.[54] ICI also had a much larger stake in maintaining the status quo, due to recent expansions in the 1970s and greater dependence on CFCs, than did American firms. What eventually convinced ICI to support CFC regulation was the prospect of two kinds of benefit from ozone regulation: first, larger profits from CFC substitutes in the long term (Maxwell and Weiner 1993, 27) and, second, the avoidance of costs of inaction. In 1986, ICI rejuvenated its research program into CFC alternatives that had been abandoned in the early 1980s—responding partly to the appearance of the ozone hole but more particularly to the renewed threat of international regulation that the hole implied and the economic pressure for international regulation coming from the United States—and by November 1988, ICI announced construction of two plants to manufacture HFCs (hydrofluorocarbons), a substitute for CFCs. Meanwhile, the new science of the scientific reports that came out after the signing of the Montreal Protocol effectively raised the costs of inaction, as did consumer-driven environmental concerns and concomitant scrutiny of the ICI corporate image, which posed a threat to sales of the entire ICI group in an industry already weakened by overcapacity. These factors pushed the European industry, including ICI, into adopting voluntary restrictions on the production and use of CFCs that went beyond the Montreal Protocol that had been agreed in 1987 (Jachtenfuchs 1990, 268).

These developments, in combination with the increasing scientific concern, caused shifts in valuation on both sides of the Atlantic. By September 1988 Great Britain joined the US EPA and West Germany in calling for massive reductions in CFCs, and by 1989 the EC had officially committed to the eventual total phase-out of CFCs. Developed countries were at last able to act as a unified pro-agreement coalition in negotiations to strengthen the Montreal Protocol.

The Developing Countries as Anti Coalition

While developed country commitments were remarkable, it was soon recognized that they would mean little if developing countries were not brought on board. The Montreal Protocol is one of the first international environmental agreements to address a truly global environmental problem in which the participation of developing countries was recognized as critical. Initially, however, developing countries had little incentive to participate in the budding ozone regime at all. The developing world consumed only 15 percent of the total CFCs consumed globally in 1986—just half the rate of consumption for the United States alone (Rowlands 1995, 167). Per capita consumption of CFCs in the North was 20 times higher than that in the developing countries

(Benedick 1998, 148–149). Indeed, only 12 developing country delegates attended the 1985 Vienna Conference. Because of the lack of active interest from the South, the Vienna Convention contains just one vague statement on the necessity of taking "into account in particular the needs of the developing countries, in promoting, directly or through competent international bodies, the development and transfer of technology and knowledge" (Article 4.2).

The problem, as developed countries began to perceive it, was one of potential increases in CFC emissions as developing countries' needs were increasing. As of 1991, the developing world's percentage of global CFC consumption had increased 6 percent in 5 years, to 21 percent. China's CFC consumption alone rose 20 percent annually in the 1980s (Rowlands 1995, 167–168) and in 1989 China announced plans for a 1000 percent increase CFC production by 2000 (Parson 2003, 164). Meanwhile, production of CFCs in the developing world doubled between 1986 and 1991 and was increasing by 7 percent to 10 percent each year (Porter and Brown 1991, 74). Developed countries therefore acknowledged the need to take measures to encourage developing country participation in the Montreal process and offered financial assistance for their participation in the negotiations. Nevertheless, developing nations were hardly visible apart from India and China (Benedick 1998, 47; Lammers 1988, 242). The United States and other developed countries then began to discuss enforcing participation in the regime through bans on trade with non-parties, and developing countries finally began to play a role in the third round of the protocol negotiations, perceiving a possible threat to access to CFCs (Miller 1995, 73) and to their plans for development, given that CFC substitutes would be much more expensive (Sell 1996, 100).

Nevertheless, developing countries continued to argue that whatever harm was caused by ozone layer depletion was due overwhelmingly to Northern industrialization. In other words, "rich man's problem, rich man's solution."[55] The developed countries were the major users and producers of CFCs and most other ozone-depleting chemicals and were the ones that would have to make changes in lifestyle and consumption patterns in order to address ozone loss. The situation was in fact one of asymmetrical deadlock created by the fact that for developing countries the costs of committing to a phase-out of CFCs were much greater than the benefits they expected to receive from protecting the ozone layer (Sell 1996, 100). Obtaining developing countries' participation in the ozone regime therefore required that they be given preferential treatment.

The Montreal Protocol of 1987 contained three mechanisms intended as inducements to developing countries. First, it allowed "Article 5 countries"—any developing country in which per capita annual consumption of controlled substances is below 0.3 kg—to delay their implementation of the Protocol for ten years (Article 5.1) and in fact increase their consumption for "basic domestic needs" up to this level over that period (Article 2A.1). Developed countries considered that this would enable developing countries to meet their legitimate needs while substitute chemicals were developed but would diminish incentives for them to become major new CFC producers and consumers. The eventually agreed quantity only represented about 25–30 percent of the existing per capita consumption in Europe and the United States and about 50–60 percent of the expected consumption level in developed countries after their promised 50 percent cutbacks over the same ten-year period (Montreal Protocol, Article 5.1; Benedick 1998, 93).

Second, the Protocol encouraged technical and financial assistance to developing countries by including commitments to "undertake to facilitate access to environmentally safe alternative substances and technology" (Article 5.2) and "to facilitate bilaterally or multilaterally the provision of subsidies, aids, credits, guarantees or insurance programmes to . . . developing countries for the use of alternative technology and for substitute products" (Article 5.3), and giving parties the right to submit a request to the secretariat for technical assistance (Article 10.2).

Finally, the Protocol prohibited trade in controlled substances with non-parties and discouraged technology transfers to them (Article 4). The trade provisions were intended to prevent competitive advantage gains by non-parties in order to stimulate participation; they constitute in effect the only enforcement mechanism in the Protocol (Benedick 1998, 91). The United States was the original proponent of the trade restrictions to strengthen the resolve of countries to meet the commitments being made; this is unsurprising given the United States' willingness to consider unilateral trade sanctions in order to induce agreement on strong ozone regulations. Winfried Lang, head of the Austrian delegation and a key actor in the negotiations, believed that the trade measure was perhaps the strongest inducement to developing countries to sign on because they would not be able to purchase or produce CFCs if they did not.[56] Indeed, by the time the Montreal Protocol was opened for signature, 8 of the 24 countries that signed on immediately were developing countries,[57] committing themselves to a 50 percent reduction in CFC production and use by 2010, ten years after the developed countries' deadline. These commitments needed ratification, but many developed country negotiators thought that most developing countries "would eventually ratify" (Benedick 1998, 99).

There were a number of flaws with the agreement from the North-South perspective, however (Rowlands 1995, 170–171). First, some Northern observers realized that even if developing countries stayed within their 0.3 kg per capita limit they could still increase their production to a point where ozone layer depletion would continue. While negotiators believed that developing countries were unlikely to expand their use of CFCs to the maximum permitted level, as technology that would soon be obsolete would be unattractive, the prospect of possible future scarcities could provide an incentive to do so.

In addition, some analysts predicted that developing countries might use their permitted increase not only to satisfy domestic demand but to manufacture products containing CFCs for export as well. This loophole could also enable a company from an industrialized country to manufacture products using CFCs in a developing country and then import the products to its home country or sell them on the world market, as neither products containing ODSs nor those made using ODSs as process agents were restricted. Developing country negotiators resisted proposals to ban the export of products containing CFCs and later did indeed try to link basic domestic needs to the need to export (Benedick 1998, 94). They argued that it was inequitable to allow Northern states to exceed their production quotas and export the excess to the South yet not allow Southern states to increase their production for export up to the 0.3 kg per capita limit, as these terms would only increase Southern dependence on the North.

Southern participation in the Protocol could not be considered a foregone conclusion in any case, despite the sticks and carrots that had been incorporated into the Protocol. The carrots of promises of technology transfer and financial assistance from the North were extremely vague, while the sticks, associated with trade restrictions against non-parties, would have no effect on countries with potentially huge domestic markets, such as China and India (Rowlands 1995, 170; Brown and Lyon 1992, 133).

Southern states were also dissatisfied with the weighted voting procedures specified in the Protocol, which were won by the United States during the 1987 negotiations. The 1985 Vienna Convention had laid down the "one-nation/one-vote" procedure that was usual in UN bodies but which was irksome for the United States. In negotiations over the Protocol, therefore, the United States had proposed a "two-step qualified majority" procedure for voting on adjustments to strengthen commitments on already regulated substances. Under this procedure a proposed adjustment would need to garner a two-thirds majority vote of the parties present, representing at least 50 percent of the total consumption of the controlled substances (Benedick 1998, 88). This voting process differed from voting on controls on *new* chemicals, which require the more standard amendment procedure entailing a simple two-thirds majority vote, the rationale behind this difference being that an amendment only enters into force after ratification by two-thirds of the parties and only binds those parties that ratify (Article 2.10). The problem with the requirement for weighted voting in the Montreal Protocol was that this language effectively gave the United States and the EC veto power. These compromises had been reached with little or no input from developing countries.

According to Ramakrishna, Southern states felt that they had been "caught napping" when the unsatisfactory provisions had been agreed, and that if another opportunity were to present itself they should react differently (1992, 152–153). Malaysia characterized the Protocol as "inequitable" (DeSombre and Kauffman 1996, 97) and by March 1989, only eight developing countries had ratified the Protocol.[58]

A major international ozone conference was held in London that spring with the goal of encouraging the developing states to sign and ratify (Rowlands 1995, 171). At this conference, however, developing countries emerged as a distinct interest group and demanded more. China called for an international ozone layer protection fund to sponsor research into alternative methods and transfer technology to those Southern countries that agreed to limit their use of CFCs. India backed up this demand, pointing to the "polluter pays" principle in the developed world.[59] Because these calls came from the two largest states in the developing world (and the two largest states overall in terms of population), they had the power to influence many other developing countries; even alone they could substantially alter the values of agreement and no agreement for the pro-agreement states. They were also helped by the Mostafa Tolba, Executive Director of UNEP, who had argued for funding to help developing countries buy substitute chemicals even before the Montreal Protocol was signed (Miller 1995, 73). This act of "claiming value" (Sebenius 1992b, 335) for the South by calling for a new fund made effective agreement costlier to achieve, just as the perceived costs of possible noncooperation by the South were also beginning to shift upwards with developing countries' increasing use of ODSs.

In the negotiated final statement of the London conference, the North admitted responsibility for the problem but was unwilling to acknowledge a need for any new structures such as a fund. The chasm between North and South grew wider at MOP-1 (Rowlands 1995, 172). The United States and some other Northern states remained opposed to the idea of an international fund, fearing that a new institution would put control of funds in the hands of the South rather than the (Northern) donor nations and asserting that this would create an inefficient and unaccountable body. Others, however, accepted the idea of a single central fund because large amounts of cash were needed. At MOP-2 more Northern states came on board with the fund concept, and the United States was increasingly isolated in its vehement resistance (DeSombre and Kauffman 1996, 98).

It was not the actual cost of the fund itself that engendered the United States' objections. Agreement on providing a fund, and on a feasible amount for the fund, became possible when the parties agreed to focus on "incremental costs" rather than "full costs": the difference between the cost of processes that use ozone-benign substances and those using ODSs. Studies by the British and by the US EPA both concluded that $200–$300 million over three years was a plausible amount for the fund (Litfin 1994, 145; Benedick 1998, 187). The parties only had to agree to fund a three-year period to begin with, although this meant that the amount of necessary funding was subject to increase for future three-year periods depending on the number of Article 5 countries ratifying the Protocol and the number of amendments ratified by each (Rowlands 1995, 174); thus, the United States' contribution, according to the standard UN assessment of 25 percent, would only amount to $40–60 million for the first period (Benedick 1998, 188), or between $8 and $25 million per year (Litfin 1994, 146)). This was settled after early debate among donor governments over the formula for burden-sharing: the United States preferred the UN assessment scale, which capped its contribution at 25 percent and which had considerable precedence, while the EC argued for a formula based on 1986 consumption, meaning that the US share would have been closer to 30 percent. The EC lost the argument because it was unwilling to provide consumption data for individual EC member countries with which to support its argument (Benedick 1998, 187).

Most administration officials in the United States accepted the idea of an international fund to assist developing countries with the incremental costs of implementing ozone-benign processes. Those who opposed it did so out of fear about the precedent that any ozone layer fund might set for future negotiations—specifically, negotiations that were just beginning on climate change. In particular, they strongly opposed the principle of additionality, or the idea that additional donor funds would be required over what donors were already giving through other mechanisms. Developing countries insisted on language specifying that ozone assistance would be "additional to other financial transfers" (Article 10.1), out of a fear that any such assistance might come at the expense of existing development aid (DeSombre and Kauffman 1996, 96). The overriding concern for US officials was the fact that climate change was a much larger problem that would therefore need many more cash inputs if the same logic of additional financial assistance were to apply there (Litfin 1994, 146; Parson 2003, 204).

Thus costs and benefits to the United States were perceived differently depending on whether climate change was added to the equation. There was also a fear in the United States that a new institution would take too much control away from the donor nations about where their contributions would be spent. This was countered by developing country suspicions of institutions being proposed to house the fund, such as the World Bank and IMF, in which voting was heavily weighted toward donors.

At a May 1990 special meeting on funding for Article 5 countries, the United States accepted the idea of a funding mechanism within the World Bank in principle, but argued against additionality. Its opposition could not be maintained, however, for several reasons. First, a representative of the United States' institution of choice to run the fund, the World Bank, stated at that meeting that the World Bank would only participate if the principle of additionality were adopted. Second, the US administration's position was supported neither by other countries (developed or developing) nor in the US Congress. Legislation was introduced in both houses to fund the US contribution in case the George H.W. Bush administration maintained its opposition.[60]

A major reason for the US eventual acquiescence in the fund, though, was that US industry did not support the government's position. While the US contribution to the new fund would be "paltry" (Litfin 1994, 146), the potential particularistic benefits that it would bring to the United States were not: US industry stood to gain potentially lucrative markets if developing countries were provided with the money to buy substitutes (Rowlands 1995, 174; DeSombre and Kauffman 1996, 94). DuPont's chairman in fact placed a call to President Bush before MOP-2 to state his support for the proposed fund (Parson 2003, 204)—a clear indication that DuPont considered itself a potential beneficiary of donor country support that would assist in developing markets for substitutes in developing countries. Then, too, it was not just American industry actors who would benefit, but the government itself. The industry estimated that potential windfall profits from markets for substitutes in developing countries would provide around $5.7 billion in taxes to the US government.[61] Finally, as Litfin points out, "if aid additionality were truly the issue, then the US could simply have chosen to cut its development assistance programs by a few million dollars" (1994, 147).

These benefits to be gained from providing a fund were juxtaposed against the real threat of no agreement on the part of developing countries, particularly China and India, if additional funds were not made part of the deal. Given the existing concern about the costs of noncooperation by developing countries, both the costs of inaction and benefits of action pointed to the economic rationality, for the United States, of providing additional funds. With this cost-benefit equation, the United States finally withdrew its objection to additionality before MOP-2, although, notably, the principle of additionality is not enforceable under the Protocol in any case (Litfin 1994, 155).

The United States took specific measures to protect its interests in the context of the new fund. First, the United States insisted on inclusion of language that the financial mechanism was "without prejudice to any future arrangements that may be developed with respect to other environmental issues" and of a "limited and unique nature" (Benedick 1998, 184). Second, the United States demanded that the fund be administered by the World Bank, a position that was vehemently opposed by developing countries despite a proviso in the US proposal that it should be overseen by an executive committee established by the parties. Third, as the largest prospective donor the United States insisted on a permanent seat on the Executive Committee to oversee the fund's budget, contributions, and disbursements. Finally, the United States wanted committee decisions to be reached through a voting mechanism weighted to reflect the size of contributions.[62]

Despite developing country objections to these conditions, the United States won substantively on the first three conditions with little compromise. With regard to the first condition, language specifying the "non-prejudicial" nature of the fund was included as Article 10.10 (despite this, it is generally accepted that the MLF did set a precedent at least for attempts to establish similar funding for other issue areas, including climate change).[63] In addition, three preambular clauses specified that (1) the ozone depletion problem was scientifically established—an implicit reference to the fact that global climate change had not yet been measured nor was likely to be conclusively documented for some time to come (Litfin 1994, 154); (2) the funds would make a real difference in overcoming the problem; and (3) the amounts needed were predictable.

On the second condition, the parties ultimately agreed that the World Bank should have a "paramount role within the fund" as "administrator and manager of the central function of the fund, financing projects and programs (Benedick 1998, 186). Two UN agencies, UNEP and the UNDP, were also given roles: research, data collection, and clearinghouse functions (UNEP) and feasibility studies and other technical assistance (UNDP). To this extent the MLF reflected the structure of the Global Environment Facility that was in its pilot phase at this time but which was rejected by developing countries as the locus of financial assistance under the Protocol. With regard to the US third demand, for a permanent seat on the Executive Committee, donor countries were divided into seven regions, each given a seat on the Committee; the United States was given its own region and thus a permanent seat.

On the United States' fourth condition, regarding voting power, developing country objections to the two-stage semi-weighted voting procedure for adjustments to commitments in the original Protocol influenced the outcome on the Fund's procedure. The South insisted on a principle of balance in the overall Protocol rather than weighted voting; the two-thirds majority of all parties would now have to include simple sub-majorities of both the Article 5 parties and the industrialized countries, giving both North and South an effective veto (Benedick 1998, 178). This principle of balance was carried over to the Fund's procedure, with developed and developing countries getting seven regional votes each on the Fund's Executive Committee.

Although it appears that the United States lost ground on this general voting question, Benedick points out that the Protocol's original two-stage "semi-weighted" voting "had in any case become moot, since Article 5 countries now had the numbers to block decisions" thanks to their combined total consumption of ODSs (1998, 177–178). In any case, donor countries as a group could not be forced by developing countries to do anything they were unwilling to do, since the effect of balanced voting is only to give equal *veto* power to both sides. In all, donor countries had little to lose. The protections the United States achieved, together with the benefits it expected to receive from contributing to the fund, made US participation in the fund cost effective from its own point of view.

There was still one more dispute to resolve. Developing countries tried to insert language into the Protocol qualifying the obligation to comply to be subject to adequate financial assistance and preferential and noncommercial transfer of technology. DuPont actively opposed the technology transfer provision, citing the need for patent protection to ensure industry's willingness to invest in development of the new chemicals. In the end, the developing countries settled for insertion of a statement acknowledging the reality that their ability to abide by the treaty would depend on the effective workings of the financial mechanism and technology transfer (Article 5.5), along with language specifying that if a developing country party believed itself unable to comply with the control measures it could notify the secretariat and it would be the responsibility of the next Meeting of the Parties to decide upon appropriate action (Article 5.4). Developing countries were still bound by their commitments, however, despite the acknowledgment that their ability to comply depends on financial and technological assistance. On the other hand, the developed countries were not bound to mandatory contributions; the voluntary nature of their contributions, or what Benedick calls a "tacit commitment" (1998, 187), was left implicit in the final text. Agreement on this point was facilitated by UNEP Executive Director Tolba's proposal promoting the concept of voluntary contributions on an assessed basis (see Article 10.6), which conveyed a sense of obligation.

The overall outcome of the intense negotiation between developed and developing countries at MOP-2 in 1990 was that China and India, the two biggest veto countries and leaders in the developing world, committed themselves to phase out certain CFCs and halons in 2010, ten years after the deadline for developed countries, and other developing countries soon followed suit. In return, Article 10, which before had simply encouraged technical assistance, was replaced by a new Article 10 setting out the terms of the new financial mechanism, to include a multilateral fund. Meanwhile, the developed countries also strengthened their Montreal Protocol commitments at MOP-2, raising the bar to full phase-out of CFCs by 2000. Two years later, at MOP-4, this phase-out would be accelerated to 1996. Developing countries did not emulate developed countries by accelerating the pace of their planned phase-out, instead keeping their deadline for CFC phase-out by 2010. However, the implementation of commitments by all countries is proceeding apace, as virtually all parties are well on their way to meeting their respective targets. There are 189 parties to the Montreal Protocol's provisions on CFCs, including developing countries that were due to make a 50 percent cut in CFCs as of 2005. Of those 189 parties, only seven were deemed to be in noncompliance with their CFC phase-out obligations as of August 2005, of which four were developing countries.[64]

US Interests and CFCs: An Analysis

The United States was willing to use the sticks and carrots that have been described here to influence other countries to support concrete, enforceable commitments on CFCs. It did so primarily through threatening trade restrictions in the first instance and agreeing to the establishment of the MLF in the second. The effective commitments to control and then ban CFCs throughout the world under the Montreal Protocol owe their existence to the fact that the United States was motivated by its own interests to do this.

To help understand why, this case may be analyzed in terms of the typology of costs and benefits of effective agreement for the United States that was set out in chapter 2. Potential benefits include environmental benefits, economic benefits from avoiding losses due to uncompetitiveness, and positive economic benefits. Potential costs of effective agreement include the cost of research, development, and marketing of any possible substitutes; the halting of economic activity affected by controls where substitution is not possible; and the cost of manipulating other actors' preferences where necessary.

Environmental benefits: The environmental benefit to be gained from placing restrictions on the production and use of ozone-depleting chemicals was initially difficult to prove. Unlike acidification of lakes or air pollution or toxins, ozone depletion could not be measured and policy-makers had to rely on the predictive capability of science, as the only data available were scientific theories and computer models. Thus, the idea that an environmental benefit would be gained was somewhat a matter of faith. Nevertheless, it was the threat of environmental damage that prompted some states and local governments to place their own restrictions on CFC use, at least in aerosols, thereby putting the burden of varying standards on American CFC manufacturers and ultimately causing them to favor standardization of controls at the national and even international levels. Continual advances in scientific knowledge, through data-gathering and assessment, confirmed and even exceeded initial fears, building a momentum for establishing international controls and for their progressive tightening.

Avoidance of economic costs: This was ultimately at least as relevant as environmental benefits for US policy-makers as it was of direct interest to important US industry actors. At the time the United States began to lead the effort for international regulation of CFCs it had already banned CFCs in nonessential aerosols and by 1984 there was a threat of litigation to enforce Clean Air Act provisions requiring legislation to ban CFCs in other uses. This threat could potentially be very damaging for US industry as it would affect an area in which the US industry was much more exposed to international competition. If the US industry had to limit its production and use of CFCs it had an interest in international regulation of CFCs in order to level the playing field with CFC industries in the rest of the world.

Positive economic benefits: Regulation of CFCs was exceptional in holding the potential of positive particularistic benefits for the same actors that were threatened with losses from the halting of CFC production. Once progress had been made on developing substitutes, particularly by powerful actors in the United States such as DuPont, the benefits that could be achieved from US leadership were substantial, and the fact that the same actors that might have lost from regulation could now expect even higher potential profits from substitutes after phase-out of CFCs meant that there was no powerful losing coalition within the United States to oppose US leadership on CFC regulation. It was a combination of these two factors—the great benefits to be gained and the fact that they would be captured by the actors potentially most affected by losses from regulation—that contributed to the achievement, and then the acceleration, of CFC phase-out.

Cost of substitutes: As for the costs of regulating CFCs, the costs of substitutes varied. For aerosols, the costs were comparatively minor; this fact contributed to the early success in banning aerosols in the United States and in some other countries in the 1970s. For more "essential" uses the costs of research, development, and marketing of substitutes was much higher and therefore their development took much longer to achieve.

Cost of halting activities: This question did not arise in the context of ozone because it was never suggested that certain activities, such as refrigeration, air conditioning, or even use of hair spray, should be halted. For CFCs, as for all the ODSs covered under the Protocol, the real question was how to develop chemicals to substitute for the ozone-depleting ones so as to permit activities affected by regulation to continue.

Cost of manipulation: The cost of bringing anti countries on board involved, first, the threat of unilateral legislation including trade sanctions against the developed countries that were unwilling to countenance the across-the-board reduction in CFCs being proposed by the United States in 1986–1987. Once the 50 percent reduction in CFCs was successfully accomplished, increasing developing country participation in the Montreal Protocol was of utmost importance. The cost of achieving this was the creation of a new fund to provide additional financial assistance to developing countries for meeting the incremental costs of compliance. In 1987, when developed countries made their first commitment to restrict CFC production, demands for funding by developing countries such as China and India were not considered seriously. Two factors ultimately overcame the costs for the United States of creating a separate fund to assist developing country implementation: the desire to avoid the potential environmental costs of increasing CFC production in developing countries, and the shift in US industry interests associated with the desire to create new developing country markets for the substitutes they were successfully developing.

Thanks to the early success of regulating CFCs, the Montreal Protocol is universally considered to be one of the most successful international environmental agreements to date. The Protocol has also been amended and adjusted over the years to regulate other ozone-depleting chemicals, contributing to its overall success. However, the stories of the negotiation of some of the other ODSs show a divergence in the degree of effective regulation obtained. Chapter 4 delves into the details of some of the more contentious negotiations over other ODSs in order to examine whether effective agreement in those areas also depended on the costs and benefits to the United States.

CHAPTER 4

Ozone Protection: The Story Continues

Once international controls on CFCs (chlorofluorocarbons) were established in 1987, attention began to turn to other ozone-depleting substances (ODSs). The numerous books and articles that have been written about the Montreal Protocol, particularly early ones, have generally paid little attention to most of the other chemicals regulated under it, but some of those negotiations have also engendered great debate and controversy. They vary in the degree to which they have been regulated, and this variation in regulatory effectiveness coincides with variation in the interests of the United States.

This chapter gives brief descriptions of the debates that have surrounded the establishment of international controls on hydrochlorofluorocarbons (HCFCs), carbon tetrachloride (CT), methyl chloroform (MC), and methyl bromide (MB): the four other ODSs, besides CFCs, that have engendered the most debate within the international community—in other words, chemicals for which different actors have differing cost-benefit assessments and for which common interests should not be assumed. An analysis summarizing the costs and benefits of effective agreement for the United States for each chemical then follows. It is seen that in general the results of the negotiations can be traced to domestic interests of the United States.

HCFCs: The Way Forward

As discussed in chapter 3, the phase-out of CFCs was closely linked to the backing of industry and the availability of alternative substances. However, while industry ultimately accepted the idea of a CFC phase-out, the question of which substitute chemicals would take their place was unclear. The main alternative to CFCs, HCFCs, entailed a heated dispute in their own right. The eventual domination of HCFCs and their treatment within the ozone negotiations provide further evidence that the United States' interests held sway, though in this case US interests did not favor as effective a solution as they did in the case of CFCs. The current and still contentious deadline for phasing out consumption of HCFCs is 2030 for developed countries (2040 for developing countries). This includes a "virtual" phase-out of 99.5 percent of consumption in developed countries by 2020, with a 0.5 percent residual servicing tail allowed for already-existing equipment. There are also trade controls on HCFCs and a freeze on HCFC production.

As mentioned in chapter 3, DuPont as well as other CFC manufacturers began to shift position in the second half of the 1980s as the production of substitutes began to be seen as commercially viable. While several substitute chemicals were developed, HCFCs came to dominate the CFC-substitute market, again thanks to DuPont (Litfin 1995, 267). It was not only DuPont that was focusing on HCFCs as a substitute chemical, as ICI (Imperial Chemical Industries) along with two other American firms had also made that decision (Maxwell and Weiner 1993, 34; Baldwin 1999, 120). However, it was DuPont, as the largest CFC producer (in rivalry with France's Atochem) and leading researcher in replacements, that was most influential in shaping the policy debates (Litfin 1994, 94; Parson 2003, 217). In 1989, the Alliance for Responsible CFC Policy (ARCFCP), of which DuPont was the most influential member, estimated that 70 percent of demand for CFCs could be replaced by improved conservation, recovery, recycling, and substances that did not release chlorine emissions, such as ammonia and carbon dioxide. Of these, hydrofluorocarbons (HFCs) were promoted to replace 10 percent of CFC demand. HFCs are characterized as less ozone-damaging than HCFCs, nontoxic, energy-efficient, and reusable, but both HFCs and carbon dioxide, contribute to climate change, of course, while other substances, such as ammonia and hydrocarbon alternatives, are faulted as being toxic or inflammable, more energy inefficient, more costly to produce, or hazardous to occupational health.[1]

For the remaining 30 percent, HCFCs were promoted as the only feasible alternatives,[2] although HCFCs, too, are a greenhouse gas. DuPont directly influenced acceptance of HCFCs, according to a Friends of the Earth (FOE) report, by "strategically renaming" the partially halogenated chlorofluorocarbons, which were formerly also called CFCs (though not regulated by the Protocol), "hydrochlorofluorocarbons", or "HCFCs".[3] DuPont also circulated a "glossy advertisement" in 1989 stating that HCFCs had ozone depletion potentials (ODPs) only 1 to 6 percent as high as those of CFCs. In addition, DuPont and other chemical manufacturers with an interest in HCFCs were able to shape the debate in their favor through their large role on the Technology Assessment Panel (Litfin 1995, 271–272).

The HCFCs were not "wonder chemicals"; they still destroy stratospheric ozone (Rowlands 1995, 119). Although DuPont emphasized HCFCs' very low ODPs, the predicted dangers of HCFCs are actually considerably higher if chlorine-loading potentials (CLPs) are measured rather than ODPs. For example, taking CFC-11 as 1, the CLP of HCFC-141b was calculated to be 0.52 over ten years while its ODP was only 0.08 (Litfin 1995, 272). Even though ODP measurements in 1987 indicated that the Montreal Protocol's 50 percent cut in CFCs would be sufficient to prevent major ozone losses, CLP calculations predicted that under these same provisions chlorine abundance in the atmosphere could reach 11 ppb—or nearly 20 times natural levels—by 2100 and still be climbing. The Scientific Assessment Panel (SAP) calculated in 1989 that even if all CFCs, CT, and MC were phased out and substitute HCFCs with average ODPs of 0.5 were employed, chlorine concentrations would still increase to 3.5 ppb by the year 2050; unless expanding HCFC use were halted by 2000, stratospheric chlorine levels would not be reduced to the 2 ppb benchmark, or pre-Antarctic "hole" level, until 2060 (UNEP 1989; Litfin 1994, 100, 131–134; Benedick 1998, 129–131; Parson 2003, 203, 167). Three years later, Solomon and Albritton (1992, 33–37) raised more questions about HCFCs, citing differences between what they called short-term and long-term ODPs. For example, five-year ODPs for HCFC-123 and HCFC-22 are 0.51 and 0.19 respectively, while over 500 years their respective ODPs are 0.02 and 0.05, because they have short atmospheric lifetimes—in other words, they dissipate from the atmosphere relatively quickly. The authors warned that some HCFCs proposed as replacements for CFCs might induce significant ozone destruction in the short term. On the other hand, because HCFCs contain hydrogen, it is thought that they are destroyed more rapidly in the lower atmosphere by natural processes than CFCs and that therefore only a small percentage of the chlorine in HCFCs affects the ozone layer (RAND Corporation 2004).

In any case, the development of HCFCs was considered to be in DuPont's interest. DuPont's CFC business was declining and the Freon Division's operating profits were well below those of the company's other divisions.[4] The fact that existing manufacturing plants for CFC-11 and CFC-12 could be retrofitted to produce HCFC-22, combined with the very much higher prices CFC replacements would garner, made HCFCs extremely attractive to DuPont. DuPont, along with ICI, promoted the traditional ODP measurements despite the fact that the discovery of the Antarctic ozone hole discredited the models that generated ODP values since they had not predicted the Antarctic phenomenon (Litfin 1995, 259; Benedick 1998, 129–131).

As ODSs, HCFCs could be regulated under the Montreal Protocol and there were numerous calls for restrictions on them (Rowlands 1995, 119). Other members of the industry, such as the two major German CFC producers, did not respond, favoring HFCs (Litfin 1994, 150; Baldwin 1999, 121). However, in 1990, DuPont and ICI united in concern that HCFCs "not be strictly regulated," although they diverged in their strategy regarding them.[5] Britain's ICI rejected any controls on HCFCs, arguing that if they were designated as controlled substances in upcoming revisions to the Protocol the chemical industry would have no incentive to invest in production of these products.[6] American industry, including DuPont, feared that uncertainty about regulation could stall investment in new HCFC products and/or tempt consumers to wait for better, nonchlorine technology (which would also encourage longer reliance on the more damaging CFCs). American producers therefore advocated long-term "time certain" phase-out dates, within the 2030–2050 period, in the revised Protocol, to reflect the 30-to-40-year lifetimes of equipment using HCFCs.

At MOP-2 some low-producing countries—Australia, Finland, New Zealand, Norway, Sweden, and Switzerland—proposed new Protocol language that would restrict most HCFCs to specified critical uses for which no alternatives existed and would phase them out completely between 2010 and 2020. Industry officials resisted this pressure—DuPont most forcefully—and asserted that, as between continuing to use CFCs and switching to HCFCs the latter was the lesser of two evils while more appropriate alternatives were developed and tested.

The United States had already established 2015 as the date for freezing domestic production of HCFCs and 2030 for phasing them out altogether in the 1990 amendments to its Clean Air Act (CAA), which were about to be made law (Shimberg 1991, 2194). The Executive Director of UNEP, Mustafa Tolba, proposed a 2040 phase-out date in the Protocol revisions. Other producer countries, particularly the EC, opposed including the compounds in the Protocol as controlled substances at all, even for reporting purposes (Benedick 1998, 145, 175). In the end, the parties compromised on a nonbinding resolution at MOP-2 that HCFCs should be phased out "no later than 2040 and, if possible, no later than 2020."[7]

By 1992, however, the European Community (EC) had regulated HCFCs within the community and now became a leader in advocating an early global consumption cap on HCFCs and a complete phase-out by 2015–2020. This effectively reversed the position of the EC and the United States at MOP-4. It was influenced by a strong Greenpeace campaign advocating and investing in HFCs rather than HCFCs for refrigeration. Opponents of HCFCs argued that suitable alternatives to both CFCs and HCFCs did exist, such as a refrigerator which was developed to use a mixture of propane and butane (two ozone-benign substances) by DKK Scharfenstein in Germany (Rowlands 1995, 120). The HCFC opponents also noted that international regulation had encouraged the development of suitable substitutes for CFCs and enhanced their marketability, while use of HCFCs could inhibit the development of better technologies that would only be profitable if HCFCs were not allowed.

Perhaps more importantly, the British chemicals giant ICI also switched its strategy to invest predominantly in HFCs and then stopped producing HCFCs altogether (Parson 2003, 182). Thereafter ICI became a "virtuous" advocate of strict controls on HCFCs,[8] enabling the United Kingdom to become a strong proponent of strict HCFC controls. Benedick suggests that these factors combined with a general desire by the Europeans to "satisfy domestic green critics by promoting an apparently strong environmental position at little cost" (1998, 205). It is not uncommon for companies to cloak themselves in a green mantle when it suits their economic self-interest. DuPont similarly "polished its environmental image" by twice advancing its CFC phase-out date before Protocol deadlines. It incurred protests from CFC users who had to bear the costs of premature equipment conversion (Parson 2003, 182), but on the other hand DuPont had an interest in encouraging the use of HCFCs, which promised large rewards to producers that could capture the market quickly.[9]

The United States was now placed in the "anti" position of opposing the most effective measures being proposed, as it demanded a residual "servicing tail" through 2030. American economics explains its position: the United States' warmer climate and concomitant reliance on industrial and commercial refrigeration and air-conditioning had galvanized much heavier investments in expensive HCFC equipment and the United States had an interest in preventing premature obsolescence of its relatively much larger stock (the fact that the United States is neither the hottest nor most populous country on earth, yet has the vast majority of HCFC-dependent equipment, did not influence the debate).

In the end, consumption of HCFCs was indeed capped as of 1996, with a 99.5 percent reduction by 2020 but the Americans won their 0.5 percent servicing tail of usage. This consumption cap was based on 1989 ODP-weighted HCFC consumption plus 3.1 percent of 1989 ODP-weighted CFC consumption. The 3.1 percent figure allowed approximately the same tonnage of replacement HCFCs to be consumed as there were tons of CFCs consumed in 1989, but this figure was later reduced to 2.8 percent. Production of HCFCs was not controlled. Controlling only consumption had two main implications: first, it allowed production in developed countries to continue to rise as long as it was used for export to developing countries; second, it meant that every new control on HCFCs would have to be ratified as an amendment, unlike controls on production of other ODSs that could be strengthened by adjustment. Rowlands attributes this victory to the "political clout of . . . industry and the US" (Rowlands 1995, 121). Developing countries, meanwhile, did not make any commitments on HCFCs or on any other chemicals at MOP-4 because of ongoing dispute over the MLF (Multilateral Fund for Implementation of Montreal Protocol).

Meanwhile, the Technology and Economic Assessment Panel (TEAP)—formed in 1990 to combine the original technological and economic assessment panels (UNEP 2006b, 149)—issued reports in 1994 and 1995 that in effect came down on the US side in concluding that HCFCs were "technically and economically necessary for effecting the conversion from CFCs" (Benedick 1998, 290–292). The 1995 report made recommendations that suited DuPont's interests directly, noting that some companies had demonstrated environmental leadership by early commercialization of HCFC technologies (Parson 2003, 228) and suggesting that parties might "consider the advantages of not turning leaders into losers by halting HCFC production too soon to allow reasonable recovery of investment costs" (UNEP 1995b, 5). The link between DuPont's interests and this language may not have been entirely coincidental; it was strongly suspected that DuPont and other chemical manufacturers with an interest in HCFCs had a prominent role in the TEAP (Litfin 1995, 272), prompting Sweden and Switzerland to propose increasing governmental control over TEAP's membership and operations shortly thereafter (Parson 2003, 232).[10]

At MOP-7 in Vienna in 1995, the Europeans officially endorsed accelerating the phase-out of HCFCs to 2015, while the United States was "the most vocal opponent to changing the status quo." The Americans argued that the European proposal would only result in a 2 percent reduction in cumulative chlorine over half a century, although the Report of the SAP for 1995 found that the total loss of ozone would be 5 percent less if hydrochlorofluorocarbons were eliminated by the year 2004 (UNEP 1995a). The Americans also predicted that accelerating the phase-out of HCFCs might cause ancillary environmental damage if uncertainty over HCFCs caused countries to delay CFC phase-out and wait for zero-ODP substitutes, or if manufacturers switched to substances with higher global-warming potential (such as HFCs) or lower energy efficiency. Underlying the American position, though, was the threat of premature scrapping of an existing worldwide capital base estimated in value at $200 billion as of 1995, most of which was located in the United States, and a threat of "disproportionately heavy costs" to the American economy (Benedick 1998, 292). In the end, the Americans held out and there were no changes to the HCFC schedule, apart from a specification that the 0.5 percent consumption allowed after 2020 is restricted to servicing of refrigeration and air-conditioning equipment in existence as of 2020 and the addition, once the MLF dispute was resolved, of a commitment for developing countries to phase out consumption of HCFCs by 2040.

The debate over HCFCs was not over in Vienna despite the fact that "the US made clear that it considered the issue now definitively closed and that continuing threats of tighter controls would be unhelpful" (Benedick 1998, 293). Instead, 24 parties issued a declaration at the end of that meeting that stressed the need to strengthen further the HCFC controls for both industrialized and Article 5 countries (UNEP 1995a), and the EU (European Union, formerly the EC) and Switzerland reopened the subject two years later at MOP-9 in Montreal. This time they were able to point to a doubling of HCFC production and consumption since 1989, accounting for up to 10 percent of remaining production and consumption of ODSs (Oberthür 1997). Both the EU and Switzerland introduced proposals that would address HCFC production. The EU introduced a proposal to accelerate the phase-out of HCFC consumption to 2015, in a stepwise process based on the reduction schedule the Europeans had committed themselves to at the regional level. It would reduce the consumption cap from 2.8 percent to 2 percent and place controls on HCFC production on the table for consideration. Switzerland, meanwhile, proposed to phase out production in industrialized countries by 2030.

A veto coalition of developing countries and the United States successfully opposed the addition of any production controls or changes to the HCFC schedule in Montreal. Developing countries suspected that limits on HCFC production in industrialized countries would reduce the availability of these substances in the South and might eventually also be applied to developing countries themselves to restrict developing country exports. As for the United States, one European observer at MOP-9 remarked that "the US appeared to have promised its HCFC producing industry to protect it from further changes of the framework conditions of their business" (Oberthür 1997, 433). This speculation is supported by a July 1997 letter from a group of industry CEOs to the State Department that refers to continuing reliance by the group on "the assurance of the US delegation" regarding US opposition to altering the HCFC production cap and phase-out timetable.[11] Thanks to the EU, however, a new declaration, this time signed by 34 parties, called for a decision on consumption and production controls on HCFCs at MOP-11.

At MOP-11 in Beijing, the EU made several proposals regarding HCFCs, some of which were ultimately incorporated into the Beijing Amendment to the Montreal Protocol. The EU again called for a 2 percent cap on consumption in industrialized countries and stronger intermediate controls. Noting that HCFCs were the only ODSs in the Montreal Protocol whose production remained entirely uncontrolled, the EU called for production controls for both industrialized and developing countries, including a full phase-out of production in industrialized countries by 2025 and for developing countries within the timeframe of their existing consumption controls. The EU also proposed a ban on trade in HCFCs with non-parties to the Beijing Amendment as a way to encourage ratifications (Oberthür 2000, 36).

The proposed strengthening of the HCFC consumption phase-out schedule for industrialized countries was again met with "fundamental opposition" from the United States (Oberthür 2000, 36). Once more it appeared that there was a US government-US industry "understanding" that no further controls restricting the use of HCFCs would be introduced, because of industry's need to have a stable basis for taking investment decisions regarding HCFCs. There was also little support from other major countries such as Japan, India or China.

However, in Beijing the EU finally had some success. The EU did not obtain stronger consumption controls or a phase-out of HCFC production, but it did obtain the inclusion of trade controls on HCFCs and a freeze on HCFC production. With regard to trade, each party to the Beijing Amendment must ban the import and export of HCFCs to and from non-parties. The details of this ban were not decided until MOP-15, however, with the result that developing countries not ratifying the Beijing Amendment will not be considered as "non-parties" to it until 2016, when HCFC production and consumption measures will go into effect for them (Alvarenga et al. 2003, 7).

With regard to production, the Beijing Amendment required industrialized parties to freeze HCFC production from 2004, with developing countries following suit in 2016. For this purpose, a party's calculated level of production of HCFCs may not exceed, annually, the average of (1) the sum, in 1989, of its calculated level of HCFC consumption and 2.8 percent of its calculated level of CFC consumption and (2) the sum, in 1989, of its calculated level of HCFC production and 2.8 percent of its calculated level of CFC production. Significantly, the fact that production controls on HCFCs were now finally included under the Protocol meant the parties could strengthen those controls in the future using the speedier adjustment procedure rather than by amendment. In the end, however, the Amendment set no deadlines for phasing out production of HCFCs. Moreover, it was also weakened by interpretative language included in the report of the meeting, which allowed both developed and developing countries an extra production allowance of 15 percent to meet the basic domestic needs of developing countries.

Overall, the EU got some of what it wanted on HCFCs in Beijing. One possible scientific reason for this success might be the fact that the ozone hole over the Arctic, discovered in 1988, was found in the late 1990s to be worse than expected (Suplee 1998). After all, that discovery that there was a *possible* Arctic ozone hole was thought to have been on the minds of negotiators of the London Amendment at MOP-2 in 1990 (Benedick 1991). However, no one close to the scene in Beijing has drawn any similar connection between results there and 1998's confirmation of the severity of the Arctic ozone hole.

There are other possible reasons for movement on HCFCs in Beijing. Oberthür (2000) has speculated that because the EU had united over an internal regulation to phase out HCFC production by 2026, it could therefore speak convincingly with one voice. This unity was enhanced by the fact that, as a result of these new EU rules the European chemicals industry now recognized an industrial standard at the global level to be in its own interest. In addition, the EU was able to leverage support from the Chinese over the HCFC issue. China had objected to the EU's proposals because of the inclusion of a freeze on HCFC production in developing countries in 2016. However, China was eager to have the parties agree on a "Beijing Declaration", a statement of general principles negotiated by consensus but not amounting to a binding treaty, in which to issue a strong call for donor parties "to maintain adequate funding and to promote the expeditious transfer of environmentally sound technologies" (UNEP 1999b). They also needed it in order to be able to demonstrate some tangible success from the conference in case there was no agreement on an amendment or adjustments.[12] According to Oberthür, this tipped the balance. With pressure from a coalition of China and the EU—two large and powerful potential rivals—the United States could not hold off consensus on some tightening of controls, especially given the anomalous character of HCFCs as the only chemical group under the Protocol that lacked controls on either production or on trade with non-parties.

On the other hand, the regulations that the United States ultimately agreed to were in fact "not really all that onerous."[13] The United States did not accept a phase-out of production, only a freeze, and as for trade with non-parties, developing country parties have no restriction on their consumption of HCFCs until 2016, so with a European ban on exports to non-parties from 2004 that had been agreed in 1998, the United States has gained the opportunity to replace those former European exports of HCFCs to developing countries with exports of its own (Parson 2003, 237). The United States' rejection of the Kyoto Protocol also means that it recognizes no controls on its HFCs—the main replacements for HCFCs—and can pursue their development freely.[14]

Compared to the inability to move HCFC controls for developed countries forward virtually at all from 1992 to 1999, the success of the EU in Beijing in finally getting agreement to freeze production and control trade in HCFCs was a major achievement. Yet the fight is still on to address HCFC production and consumption. At MOP-12 in 2000, the EU pressed developing countries on adjusting their timeline for control of HCFCs, proposing to move up the freeze on Article 5 parties' consumption from 2016 to 2007 (based on 2006 levels). The proposal also included reductions of 35 percent by 2014, 65 percent by 2020, 90 percent by 2025, and 99.5 percent by 2030, retaining the 2040 phase-out date but paralleling the developed countries' ten-year 0.5 percent consumption "tail". This proposal was watered down into a draft decision to request TEAP assessment of this issue, to be taken up at an intersessional working group meeting, with the intention for such an assessment to contribute to full MOP consideration of adjustments in 2003.

The draft decision for the TEAP assessment was presented for consideration at MOP-13, in 2001, with the EU once again taking the lead and noting a further tightening of its own regulations regarding HCFCs toward EU phase-out by 2010. Although the United States and other developed countries expressed support for the proposal, many developing countries, including India and China—large players by anyone's definition—noted the need for additional finance if an accelerated HCFC phase-out schedule were adopted, while others opposed the proposal altogether. This despite the fact that by now reports on the development of numerous alternatives to HCFCs, including HFCs, carbon dioxide, and ammonia, were becoming commonplace. The decision was eventually adopted and the TEAP HCFC Task Force delivered its report in 2003 (UNEP 2003b). Thus far it has not resulted in tightening of controls on HCFCs for Article 5 countries. The United States' expressed support for tightening controls on HCFC use in developing countries comes into some conflict with any interest in maintaining a market for HCFC exports in developing countries. The 35 percent cut on consumption of all HCFCs in developed countries came into effect in 2004; for some HCFCs this creates a significant surplus for export, given that there are no controls at all on developing countries until 2016 and even then only a freeze on consumption and production until 2040.

HCFCs were originally hailed as playing a key role in the Montreal Protocol's reputation for effectiveness, which was gained because of its rapid success on CFCs. Not only did all parties commit themselves to a complete phase-out of that chemical, the timeline for phase-out was then also quickened. This success, however, was dependent on acceptance of HCFCs as a transitional substance. With a virtual phase-out of consumption not due until 2020 for developed countries (2030 for the last 0.5 percent) and 2040 for developing countries, and no reductions in production yet required, the treatment of HCFCs may be regarded as one of the less effective elements of the Protocol. The fact that its own interests prevented the United States from leading the call for stronger HCFC controls, thanks to its relative dependence upon them, has resulted in the comparatively low level of controls obtained on this chemical.

In the United States, consumption of HCFCs was capped in 1996 but actual consumption did not approach the capped level until 1998–1999. This prompted the EPA (Environmental Protection Agency) to formulate rules for an allowance system, which were promulgated in January 2003 (USEPA 2003). The cap on consumption, though, says nothing about exports. As compliance even at the best of times can only be expected to be as good as the language of the agreement, this is not an issue of non-implementation or non-enforcement, but it is an issue that pertains to the effectiveness of the agreement.

HCFCs and US Interests

The story of HCFCs is unique in that their use was first promoted in order to substitute for CFCs but they were then subjected to calls for restrictions themselves. The United States did not lead the fight for early phase-out and tight restrictions on HCFCs, and restrictions on HCFCs are less effective than those for some other chemicals. These facts can be linked to the United States' perceived costs and benefits from more effective action during the course of establishing and then tightening controls on HCFCs.

Environmental Benefits: Initially, the environmental benefits of using HCFCs rather than CFCs were estimated in orders of magnitude. Because HCFCs were so much less damaging to the ozone layer, at least according to measurements of ODP, the benefit to be gained from the avoidance of environmental damage by restricting these chemicals was considered relatively minimal. The lesser effectiveness of the restrictions placed upon them, particularly in the length of the timeline for their phase-out, was at least partly due to this perception that there were comparatively few environmental benefits to be gained from greater effectiveness. Notably, though, the fact that the EU was able to win some of its demands on HCFCs in 1999 may have been due to increasing concern about environmental damage, given the finding that the Arctic hole was worse than expected in the late 1990s.

Avoidance of economic costs: There is no economic cost to avoid for the United States from international regulation of HCFCs as US law is not the catalyst for this action.

Positive economic benefits: There was originally a huge positive benefit to be gained from *promoting* the use of HCFCs as substitutes for CFCs, as US industry actors stood to capture new markets for these profitable substances that they took the lead in developing. As for substitutes for HCFCs themselves, there was no guarantee that any substitutes would bring the return on investment that HCFCs originally brought, so there has not been the positive economic incentive to ban HCFCs that there was for banning CFCs.[15]

Cost of substitutes: As described above, investment in research and development for substitutes for CFCs was a substantial and lengthy commitment, which is also a factor hindering more effective agreement on HCFCs themselves. Further research for better substitutes is ongoing, although none are perfect and some of them, including the popular HFCs, also contribute to climate change. The cap on HCFC consumption and production that is now in place can be linked to the movement toward other substitutes by manufacturers such as DuPont[16] and to studies indicating that production of HCFCs is dropping worldwide while that of HFCs is rising (RAND 2004). Given the ongoing research into substitutes, thanks to the incentives that do exist, as manufacturers make the shift to other substitutes it is not inconceivable that US opposition to strengthening restrictions on HCFCs might at some future point weaken sufficiently to achieve a tightening of these, should the Europeans again advance such a proposal. After all, the argument made now—that pulling the phase-out forward or setting interim goals to reduce production would cause premature obsolescence of HCFC-reliant equipment—did not hold much sway for DuPont in the 1990s when its own moving CFC phase-out dates made CFC equipment prematurely obsolescent.

Cost of halting activities: The US industry asserts the spectre of a huge potential loss from any attempts to tighten the timeline for phase-out of HCFCs, given the problems with other substitutes. As mentioned above, this is mainly connected with the loss of use of HCFC-dependent machinery and equipment if restrictions on HCFCs should make them prematurely obsolete; thus the United States' insistence on the 0.5 percent servicing tail of production of HCFCs until 2030 for developed countries.

Cost of manipulation: The cost of manipulating tighter developing country commitments to phase out HCFCs would be considerable, given the investments in HCFCs that they were encouraged to make. Given the perceived lack of benefits to be obtained and the costs to be incurred from tighter restrictions on HCFCs, the United States has no interest in manipulating developing countries' preferences to favor an accelerated timeline.

Methyl Chloroform

Methyl chloroform was a popular industrial solvent that, although identified as an ozone-depleting chemical in the 1970s (Parson 2003, 200), was only recognized as a major source of stratospheric chlorine in 1988 (Benedick 1998, 112). The case of international regulation and phase-out of MC is a success story that was not due to leadership from the United States. The United States favored weaker regulation than some states and this is one of only a few cases in which it was seemingly pulled into an effective agreement by "pro" countries favoring complete phase-out. In reality, domestic shifts in the costs and benefits of effective agreement on MC heavily influenced the US to shift its position eventually to favor international MC controls.

The solvent MC has only 10 percent as much ozone-depleting potential as CFCs, and a short atmospheric lifetime, and in the rush to halt production of CFCs after 1987 it was initially considered the best alternative to CFC-113, one of the worst ozone-depleting chemicals (Parson 2003, 200). However, the 1989 report of the SAP showed that huge quantities of MC were being produced—as much as all CFCs combined—making MC responsible for 16 percent of total industrially produced chlorine. According to Parson, MC actually represented 22 percent of all excess chlorine in the atmosphere (2003, 161) and its production had quadrupled from 1970 to 1988 with no reversal in sight (2003, 200–202). The SAP concluded that returning the Antarctic ozone layer to more natural levels would require a full phase-out of MC, among other things, but MC was not controlled at all under the Protocol. On the other hand, because of its short atmospheric lifetime, phasing out MC presented a golden opportunity to reduce short-term peak chlorine atmospheric concentration. The 1989 Synthesis Report, meanwhile, stated that it was technically feasible to phase out at least 90 percent of MC by 2000 (UNEP 1989); this figure increased to 100 percent later that year (Parson 2003, 202) as many nonfluorocarbon alternatives were becoming available for CFC-113.

Despite all the scientific evidence on the potential harm from MCs and reports on the technical feasibility of MC phase-out, this is not an area in which the United States led the push for strong controls. Even though Benedick says it was the United States that first called for a freeze in use of MC, in negotiations leading up to MOP-2 in 1990, Parson states that the

Canadian delegation had proposed limits on MC as part of its proposal for the original Montreal Protocol negotiations in 1986 (2003, 129). In fact, all developed country governments favored some regulation, but they differed on the timeline. The Soviet Union led the push for the strictest MC controls, calling for a 2002 phase-out, while some other OECD countries called for a phase-out in 2000 with a 50 percent cut by 1994 (Parson 2003, 202).

The United States' proposal, on the other hand, focused on future assessments, calling for a reduction of at least 25 percent depending on the results of such assessments. This reflected the interests of American industry by and large, which produced over half of the total amount of this chemical in the world, with Dow Chemical alone responsible for 40 percent of world production. The American MC industry argued that the maximum feasible MC controls would be a 20–25 percent cut in emissions by 2000, that substitutes were in fact not as readily available as the Synthesis Report argued and that many firms were relying on MC as a substitute for CFC-113. The MC industry also tried to assert that production was decreasing. The TEAP undermined the latter claim with data showing that worldwide sales of MC had grown by 30 percent from 1982 to 1988; indeed, an industry survey showed a 5.8 percent increase in 1989 alone (Parson 2003, 202). The SAP, meanwhile, stated that even a 20 percent reduction would increase chlorine concentrations by as much as 0.5 ppb and delay ozone recovery by 30 years (Litfin 1994, 140–141; Benedick 1998, 173).

By MOP-2 itself the United States had come as far as accepting a 50 percent cut by 2000, in a coalition with the EC and Japan. Two factors softened the Untied States position toward acceptance of a 2005 phase-out date at MOP-2. First, the negotiations over MC happened to coincide with discussions for the MLF. The Norwegians were able to link the two to apply pressure on the United States by withholding approval of preambular paragraphs concerning the nonprecedential nature of the MLF that the United States was insisting on until the United States agreed to a 2005 phase-out (Benedick 1998, 174). John Sununu, White House Chief of Staff for President Bush, was much more concerned about US safeguards on the ozone fund than about the phase-out date for MC and immediately accepted the trade-off (Litfin 1994, 151).

Second, and more importantly, the US position during the 1990 negotiations was undercut by congressional action approving amendments to the US Clean Air Act that included MC phase-out dates of 2000 (Senate) or 2005 (House of Representatives) (Shimberg 1991, 2192). Congress had overcome industry opposition and achieved agreements on MC phase-out thanks to the fact that American industry found itself between a rock and a hard place in 1990. The framers of the 1977 CAA had set in writing that its standards for clean air, specifically with regard to tropospheric ozone and particulate matter, were to be met by 1987. When numerous cities across the United States failed to meet that deadline, NGOs brought lawsuits against the EPA, and won. This posed a significant threat because if the EPA complied with the legal judgments against it, industry would be brought to a standstill. The only way out, for both the EPA and US industry, was to change the law. The Democratic Congress was able to use this opportunity to leverage industry acquiescence on new environmental controls on emissions trading in sulfur dioxide[17] as well as a domestic phase-out schedule for methyl chloroform in the 1990 amendments to the CAA.

Notably, the ten-year deadline in the CAA 1977 has been said to represent not so much any (unrealistic) scientific expectations of the time but, rather, a lack of political will to make tough decisions. Giving cities and other affected parties ten years to meet the deadline was a way of avoiding the need to address the issue sooner, aided by a hope that perhaps in ten years such a deadline could in fact be met.[18] This practice of "avoidance behavior" through setting deadlines at a seemingly distant point in the future is a common strategy for achieving agreement. Parson states, for instance, that in the case of international CFC controls, the ten-year grace period given to developing countries for meeting their obligations was "a cheap, simple, and hastily conceived response, largely equivalent to delaying consideration of the question," that was later shown to be "misconceived" (2003, 146).

With the CAA 1990, US industry now faced the prospect that its competition would have less stringent international regulation than US industry would have domestically unless a phase-out of 2005 were accepted in London. This fact encouraged industry's capitulation and the United States' shift of position, and the US delegation now lobbied for a complete phase-out. The EC was brought on board when the United States ceased its opposition to an amendment to allow EC members to transfer production facilities for controlled substances within the new European single market. Japan then came in thanks to unwillingness to block consensus, despite having a large MC industry watching its actions (Benedick 1998, 174). The developed country parties at MOP-2 thus agreed to a freeze on MC in 1993 and cuts of 30 percent in 1995, 70 percent in 2000, and 100 percent in 2005, with a review of these commitments in 1992.

Before that review could take place, a Senate resolution calling for phase-out of *all* ozone-depleting substances "as fast as possible" passed by 96 to 0 in February 1992 and the White House announced a proposal to advance existing phase-out commitments under the Montreal Protocol to the end of 1995. These actions in the United States took place immediately in response to threats by three environmental organizations to bring suit against the EPA under the precautionary obligations of the CAA if existing phase-outs were not advanced (Benedick 1998, 197; Parson 2003, 215). However, the explicit shift in US policy also occurred after a US Arctic expedition led scientists to announce that the risk of large losses of ozone in the Northern Hemisphere was greater than had been believed and that extreme ozone losses might in fact occur that very spring (Parson 2003, 215). Thus, expected costs and benefits for the United States shifted in favor of stricter controls, and agreement was easily reached that summer among all developed countries to advance the phase-out of MC, as well as of CFCs and carbon tetrachloride, to 1996.

In the end, the story of MC commitments for developed countries does not actually refute the theory that willingness and power must align in the lead state for effective agreement to result. It suggests that the United States' own economic interests might occasionally create a point of leverage that another state can exploit in negotiations, as happened in 1990. But the net benefits to the United States of international MC regulation also increased due to domestic factors, with benefits clearly outweighing the costs after domestic regulation was passed in 1990, and again in 1992, when the costs of inaction rose with scientific predictions of increased environmental harm and the threat of environmental lawsuits against the EPA.

Developing Countries, Methyl Chloroform, Carbon Tetrachloride, and the MLF

Meanwhile, developing country commitments on MC—and on CT, an otherwise uncontroversial chemical—became enmeshed in the continuing evolution of the Multilateral Fund. The Scientific Assessment Panel first brought CT under scrutiny in 1989. It found that it was responsible for 16–17 percent of total anthropogenic chlorine loading and was comparable to CFCs in its long atmospheric lifetime. It was even worse in terms of ozone-depleting potential; moreover, its global consumption in 1986 was greater in tonnage than that of all CFCs and halons combined. The reason it had not been regulated earlier was because of an oversight underestimating its worldwide usage; but it was also considered almost impossible to control because it was cheap, easy to produce, and emitted from innumerable sources (Benedick 1998, 121).

By 1990 the United States, along with Western Europe and Japan, had already severely restricted their domestic use of CT due to its serious toxic and carcinogenic properties. At MOP-2 that summer, therefore, developed countries were able to commit relatively painlessly to a full phase-out by 2000. Use of CT was much more widespread in the developing world and Eastern Europe; nevertheless, the London MOP's agreement on funding for developing country commitments brought forth a developing country commitment to phase out CT by 2010, along with a similar phase-out for MC by 2015.

At MOP-4 two years later, developed countries advanced their phase-outs of CT, along with MC and indeed CFCs, to 1996. Meanwhile, however, controversy grew over the MLF and developing countries' commitments under the Protocol. There were two issues. The first pertained to whether the MLF should be maintained at all, given the existence of the Global Environment Facility (GEF). Because the London Amendment establishing the MLF could not enter into force until 1992, an "interim financial mechanism" had been adopted in 1990, to begin functioning in 1991 (Benedick 1998, 186) and negotiations over the permanent fund continued into MOP-4 in Copenhagen. Some donors—not the United States this time—began to favor transferring responsibilities for funding on ozone to the GEF, citing administrative and economic efficiency.

The GEF was created by the World Bank in November 1990, to help developing countries fund the incremental part of the global benefits attained through meeting international treaty obligations on climate change, loss of biodiversity, pollution of international waters, and ozone depletion. The GEF is composed of the same three organizations as the MLF but its voting structure was at that time weighted in favor of donors (Rowlands 1995, 179–180). Because the MLF exists for ozone, GEF funding was shifted to provide grants to phase-out projects in "countries with economies in transition" (Eastern European countries) that are ineligible for MLF aid (Benedick 1998, 224).

The second, and more fundamental, issue than the GEF question was the fact that the MLF was not adequately financed to meet the costs of the 1990 agreements, much less any tighter controls for developing countries. Developing countries could reasonably suspect that any stricter controls for them would require self-funding for the additional costs (Parson 2003, 221). They responded to the apparent lack of donor will by refusing to take any new commitments at MOP-4 or to advance their ten-year grace periods on CFCs, MC, and CT to correspond to the accelerated phase-out schedule of developed countries. Instead, as part of the Copenhagen Amendment, they inserted into Article 5.1 a clause specifying that any new commitments by Article 5 parties would apply only after a 1995 special review of the situation that had been called for in the London Amendment of 1990 (Article 5.8), "including the effective implementation of financial cooperation" and technology transfer and the adoption of "such revisions that may be deemed necessary" regarding the developing countries' control schedule.

Although the language itself is neutral, Benedick notes that the attitudes prevailing in Copenhagen threw into uncertainty exactly what the Article 5 countries' commitments were or whether they had in fact been weakened.

However, he also acknowledges that the language had the intended effect of influencing "donor Parties to maintain interest in the 'effective' functioning of financial assistance and technology transfer" (1998, 213).

After the rancorous debate in Copenhagen, the permanent MLF was established, to take effect in 1993. However, the damaging debate produced language that left hanging the question of continued funding into the second funding period. Funding was approved at MOP-5 in 1993 for the next triennium of the MLF (Benedick 1998, 254). However, a wave of "revisionist threats" then threatened the Protocol in the United States (Benedick 1998, 226). Republicans took control of Congress in the 1994 US elections and promoted an agenda of environmental deregulation, while statements in the House and Senate advocated elimination of US support of the MLF in an effort to balance the federal budget (Benedick 1998, 295). This time, these attempts to discredit environmental science did not have the backing of US industry, as virtually all of American industry in the form of the Alliance for Responsible Atmospheric Policy (ARAP, formerly the ARCFCP) now supported both the ozone science and the Protocol's control measures. Proposed legislation to deregulate ozone-depleting substances thus "eventually fizzled" (Benedick 1998, 228) and US funding for the MLF was maintained.

While this outcome thus supports the notion that ideology alone will not hold sway in light of conflicting cost-benefit analyses by influential actors at the national level, the rhetoric emanating from the United States during this period increased mistrust on the part of both developed and developing countries. The result was that at the 1995 review developing countries again opposed strengthening their controls. They ultimately relented, but the "traditional" ten-year grace period was no longer acceptable. A few chemicals received individual phase-out commitments, but for developing countries most chemicals remained on the phase-out schedule that had been agreed in London. Any assumption the North had had that a strengthening of its commitments would always "trigger" a parallel "plus ten" strengthening in Southern commitments was exposed as wishful thinking (Benedick 1998, 297).

Benedick claims that developing country unwillingness to follow the tradition of the ten-year grace period "can be ascribed solely to MLF replenishment politics" (1998, 300); in other words, they held out for a longer grace period in hopes of pulling more money into the MLF in return for an eventual concession. His argument is supported by various analyses done at that time that showed that most developing countries would in fact phase out ODSs in advance of the London grace periods; still-existing barriers were mainly informational and administrative and were being rapidly overcome (French 1997, 168). Yet Benedick acknowledges that the donor countries now insisted on examining trade-offs between environmental impact and economic cost more closely and weighed options in terms of the MLF bill that would be presented to them. Thus, in apparent frustration of developing country hopes, "the US and other major contributors tacitly agreed with the 2010 phase-out because the estimated cost burden for the MLF was about half that of a 2006 phase-out, while the impact on chlorine abundance was not considered significant—a difference in cumulative chlorine-equivalent loading of only 2.1% over the next 50 years" (1998, 299).

The United States and the other Northern countries did not get the most effective outcome from the developing countries on MC and CT because developed countries conceded on longer phase-out times, allowing developing countries to maintain their commitments to phase out CT and MC in 2010 and 2015 respectively. As Benedick makes clear, the benefit of greater effectiveness was not perceived to merit the extra expenditure it would entail. In this case, it was in the United States' interest not to expend more resources attempting to achieve more effective developing country commitments. Unfortunately, one outcome of this lack of interest in funding was that by 1995 funding was so inadequate in general that some countries were asked to slow down planned accelerations in their phase-outs of basic substances (Parson 2003, 231).

MC, CT, and US Interests

Methyl chloroform and carbon tetrachloride were both subject to relatively little dispute among developed countries. Developing country commitments on these chemicals, however, became enmeshed in dispute over commitments surrounding the MLF. Because the disputes surrounding them involved very similar factors these two chemicals are considered together.

Environmental benefits: The environmental benefits to be gained by halting use and production were great for both of these chemicals, as they each accounted for 16 percent of anthropogenic chlorine loading.

Avoidance of economic costs: Putting international restrictions on MC and CT held some possibility of potential benefits for US industry in enabling it to avoid a loss of competitiveness due to existing stringent US legislation over these chemicals. However, incentives varied greatly for the United States with regard to this type of potential benefit. In the case of CT, the United States had virtually phased out its use earlier due to its toxic and carcinogenic properties, but because other developed countries had done likewise the United States was not confronted with a competitive disadvantage vis-à-vis these actors. In the case of MC, the United States was not in a vulnerable position as international negotiations began and the US industry successfully influenced a weak US negotiating position (Benedick, 174). The situation changed with the passage of domestic legislation phasing out MC in the United States by 2005. Rather than allow the international treaty to offer its developed country competitors less stringent controls than it faced domestically, the US industry then conceded on tighter international regulations. Competition from developing countries did not play a significant role in the United States' cost-benefit equation on either MC or CT.

Positive economic benefits: Neither the CT case nor the MC case shows any potential positive economic benefit to be obtained from international regulation by the United States or US industry.

Cost of substitutes: The cost of research, development, and marketing of substitutes did not sway the United States with regard to either MC or CT. Arguments that it was technically feasible to phase out at least 90 percent of MC by 2000, were countered by American industry officials. However, the deciding factor was that despite any potential costs of substitutes, the industry could not balk too much at the phase-out deadline set in the 1990 CAA, given that this legislation contained provisions that mattered to industry—and more powerful industrial sectors—far more than the costs of developing MC substitutes. The cost of substitutes for CT was even less of an issue, as it had already been addressed in countries such as the United States where it had been virtually phased out.

Cost of halting activities: Because MC and CT were already being phased out domestically there were no additional costs imposed on the US economy by international agreements covering them.

Cost of manipulation: As noted above, the developing countries' bid to hold out on phase-out dates in hopes of extracting more resources for the MLF did not succeed. Once adequate developed country commitments were achieved, the costs of manipulating the preferences of developing countries toward their own faster phase-out was simply considered too expensive for the amount of environmental benefit to be gained by an earlier phase-out.

Methyl Bromide

Methyl bromide (MB) was identified for inclusion on the list of controlled substances after MOP-2, becoming the most disputed of all ODSs as early as MOP-4 in Copenhagen. This chemical is used primarily as a pesticide, particularly for sterilizing soils for specialty crops such as strawberries. After being overlooked in the original Montreal Protocol negotiations, MB was estimated to account for 5 to 10 percent of observed ozone loss in 1991, although there was still much uncertainty about volume of MB emissions and its impact on the ozone layer (UNEP 1989). Although MB was known to be toxic, it was popular in the United States, where it had replaced another pesticide that had been banned for toxicity in 1984 (Parson 2003, 211). By 1991, its use was growing at a historical rate of 5 to 6 percent annually and there was no universal substitute for all the uses and pests it covered (Benedick 1998, 207). Today, however, commitments on MB include a 2005 phase-out in developed countries and a commitment to the ten-year grace period rule by developing countries, with their phase-out in 2015.

The story of MB has taken twists and turns, with the US shifting from the lead role in pushing for early bans to more recently calling for greater exemptions for critical uses as implementation concerns become paramount. In the beginning, the United States in this dispute favored effective agreement and played the role of leader of the pro coalition against the EC, despite the fact that it was the world's largest producer and consumer of MB. The reason? Arguably, the United States' immediate call for an MB phase-out just eight years down the road, and just after MB came on the international radar for the very first time, was prompted by the 1991 assessment of the SAP. The SAP concluded that MB controls would bring an opportunity to reduce ozone loss quickly: each 10 percent reduction in MB emissions might help ozone layer recovery as much as advancing CFC phase-out by three years (Parson 2003, 211).

However, it appears that in the case of MB the science was once again not the whole story. Under the US Clean Air Act of 1990, the EPA was required to ban production and import of any substance seven years after confirmation that its ODP exceeded 0.2 (in other words, 2/10 that of CFCs); methyl bromide's ODP is 0.7 (Litfin 1994, 167; Benedick, 207–208). It was thus scheduled to be banned in the United States in 2001, which would put American business and farmers at a competitive disadvantage internationally. This put pressure on the United States to avoid these economic costs. At MOP-4 in 1992, the United States therefore proposed a complete phase-out by 2000. (In fact, the United States' own agricultural interests opposed the United States' official negotiating position, and they chose to pursue a different method of leveling the playing field: lobbying against the prospective US ban itself. This position on the part of US economic actors would ultimately affect US implementation of its international commitments on MB.)

Notably, the SAP's strong 1991 recommendation on MB controls only appeared in a section summarizing implications for policy formulation. The language of the assessment report itself was weaker and more general. Parson judges that this suggests "the conclusion was formulated after the bulk of the panel's work was completed" (2003, 212). But was it politically motivated?

Tilting the executive summary of a scientific report for political ends is certainly not unheard of in the context of ozone regulation. In a case related to domestic CFC regulation in the 1970s, the executive summary of a report by the US Climatic Impact Assessment Program (CIAP) that evaluated the effects of SST operations on ozone was criticized as being "strongly influenced by political pressure" to prevent the body of the CIAP report itself "from creating a climate in which it would not be possible to grant a license to land the Concorde."[19] The SAP report may well have already been intertwined with United States' interests, as expressed at that time by the EPA, in noting a "high payoff opportunity" to reduce ozone depletion through controlling MB (Parson 2003, 211).

Other countries did not take up the 1991 SAP assessment summary's conclusion and the United States got very little favorable response to its proposal for a 2000 phase-out at MOP-4. A coalition of Israel (the third-largest producer of MB), Kenya, and South Africa mobilized developing countries to oppose even listing MB for further study. Then France, the only other MB producer, and the agricultural members of the EC who stood to lose a perceived valuable farming resource, got the EC's support. All that could be achieved in Copenhagen were a 1995 production and consumption freeze for developed countries, along with a resolution that the parties would decide by 1995 on reductions and a possible phase-out date. This was mitigated by language, originally proposed by Mostafa Tolba, the Executive Director of UNEP, that calculated levels of production and consumption would not include the amounts used for quarantine and preshipment (QPS) applications (Parson 2003, 218).

However, methyl bromide became the subject of EC regulation in 1994. The EC was split, with France and the southern European agricultural countries opposing any regulation beyond Copenhagen's freeze. Nevertheless, most members supported reductions, such as the Netherlands, which had already set a precedent by eliminating use of MB due to concern over its toxicity in soil and groundwater. However, the 1994 assessment of the TEAP MB Technical Options Committee (TOC) this time did not help the US government's case. It changed direction somewhat from that taken in the 1991 SAP, concluding that a 2001 phase-out was technically infeasible. The report was highly contentious, with the TOC divided into two factions on the issue of alternatives. In the end, it reached the conclusion that there were no technically feasible alternatives for around 10 percent of 1991 MB uses (UNEP 1994, 3).

There were reports that industry members were obstructive; industry members themselves, on the other hand, attacked the report when it was released, asserting that there were no alternatives to MB (Parson 2003, 228).

Unlike CFC producers who had sat on the original Technical Assessment Panel in 1989, producers of MB were not engaged in developing substitutes to market, and, as a result, had little interest in a reaching a conclusion calling for strong restrictions on MB use. The behavior of MB producers in the 1994 MB TOC prompted a reform of that group shortly thereafter. According to Parson, the panel did recommend that parties consider relying on an essential-use exemption rather than the existing broad exemption by usage class, as in the case of QPS (Parson 2003, 228). The critical use exemption (CUE) was later agreed—but the QPS exemption remained as well.

At MOP-7, in 1995, the European Union was split between members supporting the US proposal for a 2001 phase-out and those unwilling to go beyond a 50 percent reduction. The story of MB became linked to that of HCFCs in Vienna, with Sweden and Switzerland, among others, attempting to link the United States' desire for stronger MB controls to their own demand for stronger controls on HCFCs. The United States, with a few EC countries and Israel, rejected that linkage and managed to win a 25 percent reduction by developed countries in the Vienna revisions to the Protocol, with final phase-out by 2010. The Israelis' turnaround to favor an industrialized-country phase-out isolated France as the only MB producer country opposing controls. Developing countries committed to a freeze in 2002. A related Decision also called for review of MB control measures in 1997.

This compromise was less effective than it might appear, however. First, QPS were already exempted from controls and, second, the United States, despite having proposed a 2001 phase-out, now insisted on exemptions for "critical agricultural uses" (Parson 2003, 230), which were approved provisionally. The fact that the United States did not achieve a global phase-out in 2001 was one reason to request such exemptions, in order to alleviate the US agricultural sector's burden of having to halt MB use a full nine years before the new global deadline, and was called for in response to domestic opposition not just from private actors but from the US Department of Agriculture.

Meanwhile, the reorganization of the TEAP and the MB TOC resulted in a recommendation for a 75 percent reduction by 2001 (for both developed and developing countries) in the 1997 TEAP report (UNEP 1997b, 123) that even scientists sitting in the reformed panel thought went too far. Twenty-two scientists who participated in the assessment in 1997 accused UN Ozone Secretariat staff of presenting an inaccurate picture of the TEAP's findings. Some of them even speculated that it was pressure from the United States, specifically the US EPA, which led to the alterations in the report, due to American fears that its farmers would be competing on an uneven playing field after the US ban came into effect unless there were a full global ban of the chemical.[20]

Parson attributes concerns expressed about the report's optimism regarding technical options more to industry interests (2003, 233), and in fact there is evidence that US industry won a later backdown: a "Corrigendum" to the 1997 report replaced language calling California strawberry growers "particularly slow to make necessary investigations and investment" and "hesitant to accept" alternatives proven to be viable elsewhere (UNEP 1997b, Part 5.3, paragraph 3) with a statement that growers in California "have asserted that options from other regions are not appropriate to their situation." The Corrigendum also noted that the growers would be providing "analysis showing significant investments in research and development for methyl bromide alternatives."[21]

The controversy continued at MOP-9 later that year. The United States again sought the earliest developed country phase-out, in 2001, but once again it was stymied. Not only did the TEAP assessment report not go far enough on a complete phase-out by 2001, the United States' own economic actors did not support the United States' position. As in the case of CFCs, despite US legislation controlling an ODS—this time an impending domestic ban on MB—domestic industry did not push for matching international regulation in order to level their playing field.

Instead, the fact that the US ban had not yet come into effect enabled US domestic actors to focus on delaying the US phase-out rather than on pushing for a 2001 phase-out for all developed countries. With American farmers, and certainly methyl bromide producers, opposing the prospective 2001 ban in the United States,[22] the United States did not insist on a 2001 phase-out date in the end (Oberthür 1997, 433). Eventually agreement was reached on an industrial country phase-out by 2005, along with a restriction on trade in MB, although this did not include any restrictions on imports of products containing or produced with MB (Oberthür 1997, 433; UNEP 1997b). A "critical use exemption" (CUE) was also fully approved, for uses whose elimination would result in "significant market disruption," as long as proven alternatives were not available and emissions were reduced through "best practices" (Decision IX/6). The playing field for US agriculture was ultimately leveled by a different means when new US legislation then pushed the US phase-out of MB back to 2005, in line with this international agreement. This appeased US farmers, at least for a while—although the fact that the phase-out deadline for developing country competitors would not occur until 2015 then became a "serious issue" for agricultural and trade interests in the United States (Demetrio 1997, 1B; Mongelluzzo 1998, 1A; Fulmer 2001, C-1).

At any rate, developing countries, who had been unwilling to make any commitment beyond a freeze on production of MB in 2002, were finally persuaded at MOP-9 to return to the ten-year grace period tradition, at least for MB, with a commitment to phase it out by 2015. This commitment was the result of developed countries' willingness to concede on financing for MB phase-out in a way that they had not on CT and MC (Oberthür 1997, 433). A comparison of the language of decisions on financing taken in 1995 and 1997 (Decision VII/4 and Decision IX/5) clearly shows stronger resolve by developed countries to ensure a timely phase-out of MB in developing countries than they had demonstrated on MC and CT. In 1995, Decision VII/4 had consisted of a general call for financial support and technology transfer, with paragraph 4

> urg[ing] Parties when taking decisions on the replenishment of the Multilateral Fund in 1996 and beyond, to allocate the necessary funds in order to ensure that countries operating under paragraph 1 of Article 5 can comply with their agreed control measure commitments.

Two years later, an entire Decision was devoted to "conditions to be met" for control measures on MB in Article 5 countries. Paragraph 1.c of Decision IX/5 specifies that

> [f]uture replenishment of the MLF should take into account the requirement to provide new and additional adequate financial and technical assistance to enable [Article 5] Parties . . . to comply with the agreed control[s] on methyl bromide.

In other words, from 1999 the replenishment of the MLF would have to take into account the additional financial needs caused by the phase-out of MB. In addition, $60 million (roughly 15 percent) of the *existing* overall resources of the Fund were dedicated to MB for the rest of the 1997–1999 funding triennium, and all MB projects were made eligible for funding from the MLF on an "immediate priority" basis.[23] Oberthür states that developing countries extracted these concessions for agreeing to a phase-out (1997, 433), as they had been unable to do for CT in 1995.

The effectiveness of MB controls was still hostage to US interests, however, as became clear at MOP-11 in 1999. The EU, now attempting to take over the mantle of leadership, proposed a freeze in methyl bromide consumption for QPS (UNEP 1999b, 10), to close the loophole in the original MB control measures of the Copenhagen Amendment that had allowed QPS—representing as much as 20 percent or more of overall MB consumption—to escape controls. The United States, among other countries, objected, and the proposed freeze was not adopted, although more detailed definitions of QPS and essential uses of methyl bromide were achieved (Depledge et al. 1999, 6). It was after this MOP that the US EPA issued a revised MB rule, in response to congressional changes to the CAA in 1998, which pushed back the US phase-out to match the international phase-out date of 2005.[24]

Meanwhile, the consumption and production of MB continued to increase in Africa and Asia, although by MOP-12 the co-chair of the MB TOC stated that alternatives should be available for over 95 percent of uses and that emissions could be reduced by 30–90 percent (Depledge et al. 2000). In addition, some participants were already beginning to speculate that the election of George W. Bush might lead to a less constructive US position on such issues as phase-out of methyl bromide. At MOP-14, in 2002, the incipient split between developing and developed countries became explicit when concerns were raised that while some developing countries were moving their MB phase-out forward some developed country parties were seeking CUEs in order to avoid the 2005 phase-out deadline for developed countries. No decisions were taken on this.

This debate over critical use exemptions heated up at MOP-15 in 2003, with nominations for exemptions by the United States and some EC countries judged excessive by some observers and perceived as blocking progress in implementing the Protocol. It was speculated that in the United States, farmers were being lobbied by big methyl bromide manufacturers and convinced either that MB was safe or that current alternatives were too risky or expensive to match the chemical's sterilizing properties. It was noted, however, that the opposition of developing countries to this "breach of faith" on the part of the United States and several other developed country parties was somewhat lukewarm. This led some observers to question whether the developing countries themselves were already intending to use the same "critical use" argument in their own bargaining over MB in the future. Indeed, pleas were already being made by Algeria and Tunisia to use methyl bromide for treating fresh dates, an export commodity (Alvarenga et al. 2003, 11).

The issue of CUEs became so heated that it could not be resolved at MOP-15. An Extraordinary MOP (Ex-MOP-1) was called in March 2004 to deal only with CUEs in MB. The CUE recommendations that were adopted there represented, for some parties, more than 30 percent of their MB use in the baseline year of 1991—the year against which all reductions in MB are to be measured (Barrios et al. 2004). The United States, for its part, obtained a far larger CUE than any other country, three times higher, in fact, than Italy, the recipient of the second highest CUE (Decision Ex.I/3 (UNEP/OzL.Pro.

ExMP/1/3), Annex II/B). The fact that this exceeded the phase-out schedule for MB, which called for developed countries to cut use by 70 percent of baseline use by 2003, provoked one delegate at MOP-16 to call this a "phase-in" of MBs rather than a phase-out (Barrios et al. 2004, 12). As the deadline for developed country phase-out passed in 2005, not only had phase-out not been achieved but the critical use exemption for the United States stood at 37.5 percent of the 1991 baseline. For 2006 the United States CUE authorization was decreased to 27 percent but then raised to 32 percent in July 2005 (USEPA 2005c, USEPA 2005d).

The implications of MB exemptions are significant, as CUEs account for as much as 34 percent of some countries' baseline of MB use in 1991 (Barrios et al. 2004, 7). Meanwhile, unspecified quantities of MB for quarantine and preshipment uses also continue to be exempted from the phase-out deadline. This exemption is also problematic, with inconsistencies in parties' interpretation of the terms, lack of formal monitoring and reporting procedures in many parties, and insufficient data for making accurate estimates of how much MB is accounted for in these uses. The TEAP estimated in 1999 that quarantine and preshipment uses amount to 20 percent of baseline use of MB globally (UNEP 1999c, 40), and there are also MB exemptions for port fumigation of imported commodities and for some emergency uses (USEPA 2005a).

Not only do these exemptions preclude effective control of a toxic and ozone-damaging chemical, they also hold implications for further controls in the future. The fact that these exemptions have allowed some developed countries to avoid full phase-out as per their 2005 deadline raises questions regarding developing countries' own commitments to full phase-out in 2015. (Barrios et al. 2004, 7). Developing countries that continue to move toward full phase-out, especially when developed countries are avoiding their own, will be risking negative effects on economies that are still largely reliant on agriculture. At Ex-MOP-2 in 2005, an EU proposal to tighten developing countries' interim deadlines for reductions of MB before 2010 was put on hold until questions on critical use nominations and exemptions for non-Article 5 parties for 2006 were resolved (Barnsley et al. 2005, 2).

The exemptions also raise concerns for farm workers as well as for producers of MB alternatives and practitioners of alternative agriculture. First, methyl bromide is a Level I toxic chemical—the most toxic among four categories set by the EPA. Symptoms of exposure include eye and skin irritation, headaches, tremors, and nausea, while higher levels of exposure have been linked to vertigo and convulsions. Thus, ironically, those who actually work the land are not necessarily those who stand to gain a net benefit from continued MB use, given the costs of their exposure to the chemical. Rather, it is agribusinesses that have benefited from the ban, or, according to some, the MB industry that has benefited from convincing MB users that there are no appropriate alternatives.[25] Indeed, it is those in the industry who have been most intransigent who stand to benefit the most. As a result of these producers' lack of preparation for the phase-out through investment in the necessary research, they can credibly claim a lack of proven alternatives. Meanwhile, producers that have put effort into developing alternatives to MB, such as alternative pesticides, are not being rewarded for their efforts (Barrios et al. 2004, 12). Perhaps this reflects the fact that they have much less influence than DuPont had at the time the TEAP noted the possible desirability of catering to producers of CFC alternatives by not eliminating HCFCs prematurely. On the other hand, it has been speculated that if exemptions were disallowed, some key parties—such as the United States—might consider withdrawing from the Protocol entirely (Alvarenga et al. 2003). As of MOP-17 in November 2005, though, this did not appear likely, as the US and the EU reached a compromise on 2007 CUEs and supplemental CUEs for 2006.

MB and US Interests

At the mid-point of the first decade of the third millennium, it was methyl bromide that engendered the greatest debate within the ozone regime. This case, like that of CFCs, appears at first glance to be another story of success on the part of the United States in pushing for an effective agreement. The United States achieved a 2005 phase-out in developed countries, with developing countries following ten years later thanks to strong language calling for additional funding through the MLF. In this, the result of the push for MB regulations is markedly different from that for MC, CT, or HCFCs.

On the other hand, the behavior of some countries—including the United States—since the 2005 phase-out date was agreed shows that even a binding, enforceable commitment to a phase-out may not be enough to constrain strong actors with other interests. Thus, paradoxically, although the United States was the first country to push for MB phase-out some 15 years ago, the same American economic interests that caused the United States to put MB on the international agenda at that time have more recently benefited from the avoidance of full phase-out through critical use exemptions. Once again, a breakdown of the United States' costs and benefits shows why this is so.

Environmental benefits: The discovery in 1991 that MB posed dangers to the ozone layer meant that its phase-out could provide significant environmental benefits. Indeed, under US law, the EPA was already required to regulate it as a dangerous substance because its ozone-depleting potential was over 0.2.

Avoidance of economic costs: Of concern to the United States was the fact that MB came under tight domestic controls under the 1990 Clean Air Act, which would require phase-out in 2005. The competitive disadvantage this gave to US industry would be a significant cost, making international regulation to level the playing field a very desirable benefit, at least as far as US government actors were concerned. The American MB industry, however, was not a strong supporter of international regulation as the CFC industry eventually became, probably because the development of substitutes for MB would not provide the same windfall that substitutes for CFCs had. In addition, the fact that the domestic regulation had not come into force yet meant that industry still had some time to fight it domestically. This would benefit industry more than international regulation, so industry took an "anti-agreement" role in opposition to United States' own government actors.

Positive economic benefits/Cost of substitutes: The identification of MB as a significant ODS gave rise to activity in researching substitutes. In 1994 already-available alternatives were identified for 90 percent of MB applications (UNEP 1994). There is little evidence that marketing substitutes would give US industry a windfall benefit, though, and so far industry has been successful in arguing that it has made a significant outlay in research and development but that no effective, economical, and environmentally friendly alternatives exist for some uses.[26] This argument directly contributed to the United States' success in receiving a critical use exemption for 2005 of 37.5 percent of baseline.

Cost of halting activities: The TEAP has consistently reported progress in developing alternatives, making the question of halting activities almost moot. However, for uses for which alternatives have not proven acceptable, this risk might exist. For instance, the question of moving strawberry production in California to different locations where pests can be controlled without MB, raised in the TEAP 1997 report (UNEP 1997b, 120), entails the cost of halting activities at least for certain locations. This cost would certainly be a factor in the United States' consideration. Rather than accept this cost as a price for an effective agreement, the actors involved have been able instead to weaken effectiveness, such as by persuading the TEAP to weaken this language in its "Corrigendum" and through the United States' later insistence on CUEs, which further weakened the agreement (UNEP 1997a).

Cost of manipulation: Developing countries represented a hostile anti coalition during the first round of negotiations over MB at MOP-4, refusing to talk about any controls at all apart from a developed country freeze on production (Benedick 1998, 208–209). The price exacted by the developing countries for agreeing to a 2015 phase-out five years later was $60 million in MLF resources for MB-related activities from 1997 to 1999, eligibility for funding for all MB projects from the MLF fund on an "immediate priority" basis, and explicit acknowledgment of the need to take account of the additional financial needs caused by the phase-out of MB in the next replenishment of the Fund. Production in developing countries was considered a significant potential environmental problem if uncontrolled usage continued, however; even more importantly, the United States was threatened with a competitive disadvantage from a unilateral domestic phase-out of MB. The developing countries' price was therefore paid. Whether developing countries will adhere to the deadline or will follow the lead of some developed countries in insisting on large critical use exemptions remains to be seen, as does any developed country response.

The evolution of commitments to control MB is a unique case of US efforts to advance tight international restrictions based on its own domestic legislation, followed by US efforts to undermine those same controls through the "critical use" loophole. More than anything this demonstrates the effects of conflicts between different domestic actors. The fact that the United States pushed its own domestically legislated deadline back from 2001 to 2005 to match what had been achieved at the international level, and even now continues to find loopholes for producing MB ultimately shows, however, that in the end money wins out.

US Interests and ODSs: Conclusions

The case of the ozone agreements provides the opportunity to compare the effectiveness of agreements for controls over various chemicals in an overall successful case. Table 4.1 highlights some of the key characteristics of CFCs, HCFCs, CT, MC, and MB, the chemicals that have been the subject of greatest divergence in interests and the greatest controversy among key state actors under the Montreal Protocol.

The cross-chemical comparison of the cases below reveals interesting similarities and divergences. Two similarities stand out. First, all five chemicals were similar in offering perceived environmental benefits from restrictions, although the benefits of restricting HCFCs were downplayed by the United States. Second, the United States and other pro countries stood to risk significant costs if they chose to attempt to manipulate the preferences of anti countries in any effort to get agreements on controls or on strengthening controls. The implication of these similarities is that neither environmental benefits nor costs of manipulation could have determined the United States' decisions on which regulations to pursue. For that, the differences between the chemicals must be examined.

Table 4.1 Characteristics of CFCs, HCFCs, CT, MC, and MB

	CFCs	*HCFCs*	*MC*	*CT*	*MB*
Developed country phase out due	MOP-2: 2000 MOP-4: 1996	MOP-7: ~2020 (0.5% "tail" for maintenance purposes until 2030)	MOP-2: 2005 MOP-4: 1996	MOP-2: 2000 MOP-4: 1996	MOP-7: 2010 MOP-9: 2005
Developing country phase out due	MOP-7: 2010 = "London + 10" grace period*	MOP-7: 2040 (freeze in 2016, ten years beyond "Copenhagen + 10" proposed grace period)	MOP-7: 2015 = "London + 10"	MOP-7: 2010 = "London +10"	MOP-9: 2015 = "Vienna + 10"
MLF funding provisions	1990: Established the Multilateral Fund to assist developing countries in meeting control provisions	No special provision	Vague	Vague	Decision IX/5: additional financial needs for phase-out to be taken into account as a "condition to be met" for fulfillment of control
US position on developing country phase-out	Committed new resources to the Multilateral Fund	1995: Tacit agreement on long grace period	1995: Tacit agreement on long grace period	1995: Tacit agreement on long grace period	1997: Agreement to "condition" that MB be funded as an MLF priority
ODP	CFC 11 = 1	HCFC-22 = 0.055 HCFC-123 = 0.02–0.06 HCFC-141b = 0.11	0.1–0.16	1.0–1.2	0.7
CLP	CFC 11 = 1	HCFC-22 = 0.26 HCFC-123 = 0.035 HCFC-141b = 0.29 (>90% probability)	0.1	1	N/A

Continued

Table 4.1 Continued

	CFCs	*HCFCs*	*MC*	*CT*	*MB*
Atmospheric lifetime (years)	CFC-11 50 years CFC-12 102 years CFC-115 1,700 years	HCFC-22 = 13 years HCFC-123 = 1.2–2.4 years HCFC-141b = 9.3 years	6.3 years	50 years	1.5 years
Extent of problem	Widely used; known as wonder gases because they are long-lived, proven to destroy the ozone layer	Widely used: known as "transitional substances" for CFCs (1989); usage decreasing in developed countries, increasing in developing countries	Widely used; responsible for 16% of anthropogenic chlorine loading (1989)	Already restricted in most developed countries; cheap and widely used in developing countries; responsible for 16% of anthropogenic chlorine loading (1989)	Wide use; industrialized countries account for 80% of consumption (1994); consumption increasing in developing world; responsible for 5–10% of total ozone loss (1991)
Substitutes available?	Yes	Yes	Yes (1989)	Yes for most uses (1989)	Yes for most uses (disputed by industry); substitutes being researched

* Although CFC use in developing countries came under international controls in 1987, the developing countries' phase-out date for CFCs, MC, and CT was not firmly established until 1995, by which time debates over the MLF at MOP-4 had prompted developing countries' mistrust regarding donor countries' commitment to the Fund. Thus developing country commitment to phase-out of these three chemicals was weaker than it could have been.

Sources:

Benedick 1998
Litfin 1994
UNEP 2004a
UNEP 2004f
UNEP 2005a
UNEP 1999d
UNEP 2003b
Pyle et al. 1992
Elkins 1999
UNEP 2005b
"Trade Names of Chemicals Containing Ozone Depleting Substances and Their Alternatives" UNEP, The OzonAction Programme Library. Web address: http://www.uneptie.org/ozonaction/library/tradenames/trade_chem.asp?id = 87 (September 17, 2005).

One noticeable difference between the chemicals lies in how developed countries addressed developing countries' demands for assistance in meeting the costs of controls. A potential benefit to the United States from leveling the playing field with its competitors was present in the CFC case, but CFCs were also unique among all the chemicals in holding the potential for positive particularistic economic gains to powerful segments of US industry. This is linked to both a commitment to rapid phase-out for developed countries as well as the original establishment of the MLF in order to provide an incentive for developing country participation in the Protocol.

For MB, too, there was an economic benefit to be gained by the United States, at least at first, from strict international regulations in order to avoid a loss of economic competitiveness. The United States had in interest in keeping the competitive playing field level with regard to its own producers of MB, given the threatened 2001 deadline for phasing out MB in the United States. Southern European countries were likewise worried about competition from some African countries. It was paramount therefore to ensure that developing countries had an incentive to agree to a phase-out. This interest facilitated a decision at MOP-9 in 1997 that additional funding should be provided through the MLF for assistance in complying with MB controls.

With regard to CT and MC, on the other hand, once developed countries moved their deadline for phase-out up to 1996, it was expected that developing countries would move theirs up as well, in keeping with the Protocol's ten-year grace period. However, the cost of providing additional funding to developing countries was perceived by donors to be greater than the additional benefit they would gain from this earlier phase-out, as the only benefit would be a relatively small environmental benefit to the ozone layer itself. The phase-out deadline for developing countries thus remains at 2010.

Despite the differences in the regulation of the various ozone-depleting chemicals under the Montreal Protocol, these differences pale when comparing the success of the Protocol to other global agreements. It is generally considered to be the most effective environmental agreement that has been negotiated at the global level, and there are good reasons for this. Taking its provisions collectively, the fact that there are indeed commitments on firm *targets*—the phase-out of the full range of ODSs—with agreed *timetables*, is a solid accomplishment by any standard. It is enhanced by the fact that countries were able to agree on language in the Protocol creating mechanisms to enforce and to assist compliance. Moreover, the ozone agreements are truly global agreements, as demonstrated by the almost universal participation of both developed and developing countries. No other global environmental treaty can compare in effectiveness with the Montreal Protocol.

Appendix 4.1: Controlled Substances Under the Montreal Protocol

The following is a list of chemicals that are controlled under the Montreal Protocol. Each set of chemicals is listed in the order in which it was included (this list is limited to the maximum commitments as they exist in 2006. The meeting at which the strongest commitment was made is indicated parenthetically. A distinction is also made for commitments taken by countries classified under Article 5 of the Montreal Protocol.

CFC-11, -12, -113, -114, -115: Beginning with a baseline cap on consumption, based on 1986 levels, countries are required to freeze consumption of these in 1989 (MP), cut consumption by 75 percent by 1994 and phase out by 1996 (MOP-4, Copenhagen, 1992). Article 5 countries are required to freeze consumption in 1999, cut by 50 percent by 2005, 85 percent by 2007, and then phase out by 2010 (MOP-7, Vienna, 1995).[27]

Halons 1211, 1301, 2402: Beginning with a freeze in 1992 (MP) based on 1986 emissions, countries are required to phase out consumption by 1994 (MOP-4). Based on a baseline of average emissions from 1995 to 1997, Article 5 countries are required to freeze consumption by 2002, cut by 50 percent by 2005 and phase out by 2010 (MOP-7).

Other fully halogenated CFCs: Countries are required to cut consumption by 20 percent by 1993, based on a 1989 baseline, by 75 percent by 1994, and phase out by 1996 (MOP-4). Article 5 countries are required to cut consumption by 20 percent in 2003 based on a baseline of average consumption over 1998 to 2000, cut 85 percent by 2007, and phase out by 2010 (MOP-7).

Carbon tetrachloride (CT): Based on a baseline of 1989 consumption, countries are required to cut consumption by 85 percent by 1995 and phase out by 1996 (MOP-4). Article 5 countries are required to cut consumption by 85 percent by 2005, based on a baseline of average consumption over the years 1998 to 2000, and phase out by 2010 (MOP-7).

Methyl chloroform (MC): Countries are required to freeze consumption by 1993, based on a 1989 baseline, cut 50 percent by 1994 and phase out by 1996 (MOP-4). Article 5 countries are required to freeze consumption in 2003, based on a baseline of 1998–2000 average consumption, then to cut by 30 percent by 2005, 70 percent by 2010, and phase out by 2015 (MOP-7).

Hydrochlorofluorocarbons (HCFCs): Based on a baseline of 1989 consumption, countries are required to freeze consumption by 1996, then cut by 35 percent by 2004, 65 percent by 2010, 90 percent by 2015, 99.5 percent by 2020, and phase out by 2030 (MOP-4). They are also required to stabilize production by 2004. Article 5 countries are required to freeze consumption by 2016 based on 2015 levels, then phase out by 2040 (MOP-7).

Hydrobromofluorocarbons (HBFCs): All countries are required to phase these out by 1996 (MOP-4; MOP-7 for Article 5 countries).

Methyl bromide (MB): Countries are required to freeze consumption in 1995, based on 1991 levels (MOP-4), cut consumption by 25 percent by 1999, 50 percent by 2001, 70 percent by 2003, and phase out by 2005 (MOP-9, Montreal, 1997). Article 5 countries are required to freeze consumption in 2002 based on average 1995–1998 consumption levels (MOP-7), cut by 20 percent by 2005, and phase out by 2015 (MOP-9).

CHAPTER 5

Unconventional Behavior on Forests

The case of forests lies at the opposite extreme of effectiveness from the case of international efforts to protect the ozone layer, as endeavors to achieve consensus on a global forest convention have repeatedly failed. Because of the failure to achieve a legally binding global agreement on forests, relatively little academic attention has been given to this process. Such failure, however, makes deforestation useful to compare with the ozone case in order to judge any theory purporting to explain the success of the ozone agreements, particularly if history is to teach us any lessons about how to address other global environmental problems such as climate change.

In many ways, the cases of forests and ozone are similar. Just as with the loss of ozone, the loss of forests came under international scrutiny in the 1980s, a period of increasing concern over global environmental problems. Just as ozone performs the environmental service of UV filtration, forests perform numerous environmental services, such as oxygenation and moderation of surface and air temperatures through utilization of carbon dioxide (their "carbon storage" values). In the case of forests there were actually even more environmental values behind global concern for their fate. These include numerous "biospheric functions" (Guppy 1996), such as absorption and reflection of solar radiation; utilization of carbon dioxide and water vapor to photosynthesize sugars and starches, oils and fats; and later, adding salts and other substances, to make proteins and other basic foods for life on earth; filtration of dust and pollutants; purification of rainwater and release of water vapor, which then creates further rainfall; and the stabilization of soils, which protects landscapes from erosion, among others. These environmental functions are in addition to the aesthetic and existence values of forests that are also felt on the global scale. It was the threat to these services, particularly from widespread deforestation in tropical forests, which attracted global attention to the problem of deforestation.

In the 1970s forests, like CFCs (chlorofluorocarbons), became the subject of US regulations that made US industry less competitive than its foreign counterparts. Once again, the United States proposed the negotiation of a global convention that could then be followed by more specific protocols. Negotiations on a forest convention began in 1990 under the auspices of the United Nations, as part of the preparatory process leading to the UN Conference on Environment and Development (UNCED) in 1992. Just as the case of ozone depletion became enmeshed in North-South politics (the Montreal Protocol being one of the first environmental treaties to be so affected) so, too, did deforestation. Unlike the case of ozone, however, this cleavage brought negotiations on a forest convention to a screeching halt, and the United States and the pro coalition were only able to achieve a nonbinding agreement at UNCED, the "Non-legally Binding Authoritative Statement of Principles for a Global Consensus on the Management, Conservation, and Sustainable Development of All Types of Forests" (the Forest Principles). In almost a decade and a half since UNCED, repeated multilateral efforts to garner consensus on renewing negotiations for a forest convention have thus far all failed.

Several analysts have grappled with the question of why no binding agreement on forests was produced in 1992. Lipschutz, for instance, claims that in large part this could be due to the fact that forests are different from other portions of the natural environment whose role is directly related to international commerce and can therefore be addressed through instruments that regulate trade (Lipschutz 2000/2001). This explanation, however, ignores the existence of binding agreements such as the Framework Convention on Climate Change and the Convention on Biological Diversity. Dimitrov, on the other hand, identifies a lack of information on possible transboundary consequences of forest degradation, particularly with respect to the effects of deforestation on climate change and biodiversity, as key to the failure of the forest negotiations in 1992 and a continuing lack of effectiveness in global forest policy-making (Dimitrov 2003). This argument is limited in that Dimitrov does not actually ascribe causality to this factor. It is consistent, however, with my argument that costs and benefits matter. The more important question, though, is whose overall costs and benefits matter most?

Beyond the arguments of Lipschutz and Dimitrov, the prevailing argument as to the reason for the failure to achieve a global forest convention in 1992 was that Malaysia and other developing countries were so single-minded about preserving sovereignty over their natural resources that this precluded agreement on any convention.[1] This explanation begs the question of why, if the United States is the dominant power in the international system, it did not have the power to overcome such an objection. I offer an alternative explanation, which attributes the outcome not to a lack of power on the part of the United States but to a lack of American willingness to perform an effective leadership role in the negotiations, even though the United States was the architect of the original proposal at the state level. This lack of willingness was directly related to US economic interests.

The Problem

International manifestation of concern about deforestation dates almost as far back as concern about the ozone layer. A meeting held during preparations for the 1972 UN Conference on the Human Environment (UNCHE) considered the environmental side-effects of the development process, including loss of forest resources from expansion of agriculture in tropical regions. However, at UNCHE itself Southern countries turned the discussion toward issues of economic development rather than environment and forest issues were completely overshadowed (Kolk 1996b, 129–130).

Environmental issues took a back seat to more immediate economic concerns in both North and South after the oil crisis of the early 1970s and it was not until the mid-1980s that deforestation once again rose to the international political agenda. It was not in developing countries with territorial jurisdiction over the most threatened forests that the threat of deforestation was felt most severely, despite the fact that it was their populations for whom forest health would presumably have more salience. Concern about deforestation, paradoxically, was much more a phenomenon in developed countries, where forests were valued from afar on a more global scale. The truly international effects of deforestation now began to receive attention, as tropical deforestation posed a threat not only to unique ecosystems of unparalleled biodiversity but also to the global climate.

The complex web of causes and consequences of tropical deforestation were just beginning to be appreciated by the end of the 1980s. A number of contributing factors were identified, including conversion of forest to agricultural and grazing land, fuelwood cutting at an unsustainable rate, commercial timber logging, atmospheric and climatic sources of deforestation, including acid rain, and side-effects from mineral exploration and production (Thomas 1992, 245–254). A US government background paper from 1990 cites as its two overriding concerns the severe and widespread temperate and boreal forest declines in eastern and western Europe and elsewhere due to acid rain from fossil fuel combustion and deforestation of tropical forests due to deliberate conversion of forest lands. The paper notes that deforestation had by 1990 already consumed half of the world's tropical forests, at a rate of 11 to 17 million hectares annually.[2] Actually, 40 percent of the world's *total* forest cover had been lost by 1980, but temperate and boreal forest countries had denuded their forests much earlier; since the late 1970s trends in those countries have been marked by increasing forest cover due to widespread reforestation efforts. Tropical forests were also singled out for concern because of their much greater biological diversity (Panjabi 1997, 104). The background paper predicted potential consequences due to forest loss as ranging from destabilization of the global climate to destruction of water supplies, erosion, and desertification.

As a result of international concern over tropical deforestation in particular, several international efforts were undertaken in the 1980s. First was the International Tropical Timber Agreement (ITTA) of 1983, negotiated under the auspices of the UN Conference on Trade and Development (UNCTAD). As a commodity agreement, the formation of the ITTA was heavily influenced by North-South politics, particularly the South's calls for a New International Economic Order (NIEO) and for stabilization of commodity prices in the 1970s. However, as a unique treaty in dealing with a special naturally occurring renewable resource, the ITTA negotiations received attention from environmental NGOs, who were able to have an influence at a time when international awareness about tropical deforestation was growing. Despite this, the ITTA and its implementing organization, the International Tropical Timber Organization (ITTO), have come under heavy criticism, particularly for being weighted toward the interests of the tropical timber industry. A second international endeavor on tropical deforestation, the Tropical Forestry Action Plan (TFAP) of the UN Food and Agriculture Organisation (FAO), followed the ITTA in 1985. Initiated in response to dissatisfaction with existing forest projects it soon suffered severe criticisms itself, emanating primarily from Northern countries.[3] Due to the lack of Northern support, the program stagnated in the 1990s and attention shifted to other initiatives and organizations. A third initiative that addressed forests was the UN General Assembly's decision to hold the UNCED in Rio de Janeiro, Brazil, in 1992.[4] As noted in chapter 1, this decision was the outcome of the rising concern about both environmental and developmental challenges facing the world in the 1980s and the nexus between them. A major focus of UNCED was to be the formulation of a global "action plan" for the twenty-first century, including chapters on numerous environmental issues as well as cross-cutting issues related to development and capacity-building for meeting those environmental challenges. One chapter was to be devoted entirely to global forest issues.

It was in this context of increasing Northern concern over environmental problems such as deforestation, a growing North-South political chasm over economic growth and development issues, and a growing realization by the South that the North's environmental concerns might provide a source of leverage to achieve developmental aims, that a global forest convention was proposed in 1990. Two years later, at UNCED, the nonbinding Forest Principles were agreed, along with other documents including the (nonbinding) global environmental action plan, entitled Agenda 21. Since then, calls have continued for negotiation of a binding global forest convention (though not by the United States); up to now, however, anti-convention forces have prevailed.

Forests and the Criteria of Effectiveness

In the area of global forest policy, because no binding treaty has been concluded the scores on all criteria equal zero, for nonexistent. In other words, there are no binding global commitments to protect the world's forests; because no commitments exist, no mechanisms are required to ensure compliance, and because no binding agreement exists there are no parties to agreement. Finally, the relationships of existing bodies and instruments that do overlap in the forest issue area are not clearly set out anywhere. The nonbinding document that was agreed, the Forest Principles, also lacks language itself that would meet any of the criteria listed.

Interests and Outcomes

There is remarkable confusion surrounding the genesis of the proposal for a world forest treaty, especially considering the small number of works that discuss this case. There is confusion over whether the earliest calls specified all types of forest or just tropical forests (Agarwal, Narain, and Sharma 1999, 227) and over the original author of the proposal. One source claims that the original proposal came from the United States and called for a ban on logging in tropical forests.[5] Some writers assert that the idea came originally came from the FAO (Kolk 1996b, 153; Kiss and Shelton 2000, 366) while others point out that in fact this mention of the idea came from former Swedish Prime Minister Ola Ullsten, chair of an independent commission that reviewed the FAO TFAP program in 1990 (Johnson 1993, 103; Agarwal, Narain, and Sharma 1999, 227).

Humphreys identifies nine separate proposals for a global forest instrument that surfaced in 1990, of which the United States' was the fifth. It was, however, the first proposal from a *government*[6]—in other words, the first official proposal from a prospective negotiating party rather than simply a recommendation from a nongovernmental body or international organization—that called for a global forest convention.[7]

Initial US Interests

Numerous factors have been cited as leading to the calls for a convention. There was great public concern in the North over tropical deforestation, but this was not the only motivation. This concern over deforestation led to calls for boycotts of tropical timber and other actions in the late 1980s. Hundreds of localities in West Germany had banned governmental use of tropical wood, the Netherlands and numerous other European and American state governments had announced their intention to halt the use of rainforest wood for building material, and the United Kingdom's Prince Charles had called for a boycott of tropical timber in the Western world (Agarwal, Narain, and Sharma 1999, 227). These actions prompted suggestions that a convention would appease those who would otherwise be motivated to take such measures (Kolk 1996b, 154). As the assistant director general of the FAO's Forestry Department pointed out in September 1990, a binding convention that reflected international recognition of both the economic and ecological importance of forests might benefit both industrialized and developing countries that were "jointly fac[ing] calls for a moratorium or ban on forest utilisation of any kind over large areas of forest" and thus would help the tropical timber trade, although some countries, particularly Malaysia, voiced a fear that one factor behind the proposal was a desire to raise even more universal trade barriers against tropical timber (Kolk 1996b, 154).

For the United States, a forest convention was expected to bring several additional potential benefits at low cost. By seeking the environmental benefit of halting forest loss, the United States would not only avoid the costs of this environmental damage but the government might also gain political benefits at low economic cost in being seen to address the problem. In the words of the lead Malaysian negotiator on forests at the time, the United States (and other developed countries) wanted to "appease their public opinion and thus get electoral mileage out of forests" (Ting Wen-Lian, quoted in Panjabi 1997, 142).

A second professed interest of the US government was to address market distortions in other countries that were seen not only as encouraging deforestation but also as subsidizing cheap timber exports from certain countries into the United States. This could potentially benefit the United States economically by raising environmental standards in other countries to be more equivalent to those in the United States. Environmental standards that affected US forests to some extent were incorporated into the Clean Air Act, the Clean Water Act, and the Endangered Species Act. In addition, the US National Forest Management Act has been called "a rare example of domestic legislation which requires the enhancement of biodiversity" by two Canadian observers.[8] A forest convention that would raise standards of forest practices in other countries would allow the US industry to avoid the competitive disadvantage that it faced due to the costs of having to meet higher standards.[9]

Notably, however, the US forest industry generally did not come out as a strong proponent for a convention.[10] Rather, the dominant feature of industry involvement in the push for a world forest convention was a claim of unfamiliarity with all the new issues raised during the UNCED process, the first international discussions ever to address temperate forests.[11] This may be linked to the fact that the industry is fragmented into numerous, frequently conflicting interests.

The segment of the industry most likely to be hurt by the costs of high US standards, and who thus would benefit most from "neutralizing the opposition," were those involved in the primary industry, the nonindustrial private forestland (NIPF) owners, who own about 60 percent of the commercially productive forestland in the United States.[12] However, the NIPF owners are less economically powerful than the processing end of the industry. Since 1947, the forest products industry has annually contributed about 6.5 percent of total manufacturing contributions to the US economy, thanks to value-added processing and timber production,[13] or about 13 times as much value added as is derived from timber harvesting on nonindustrial private forestland. Moreover, there is no organization representing the interests of these less powerful actors against those of the industrial giants. In the words of one US official:

> Small industry is not organized; other than the state foresters and a few associations there's no one that represents them, but those associations not only have Joe Blow, they probably also have Weyerhauser, and their livelihood doesn't depend upon Joe Blow, it depends upon Weyerhauser.[14]

As a result, the actors who would be most benefited by raising standards internationally had comparatively little voice in the development of the industry's position.

The wood processing industry is not as hurt by US environmental restrictions as are the forest landowners. The costs of meeting forest management standards, as reflected in the price of the raw timber, constitute only a small proportion of the final price of the processed wood product. In addition, although the industry claims an interest in buying locally—due to the cumbersome nature of transporting raw logs—US wood processors can and do sometimes benefit from the ability to buy raw logs from overseas rather than the more expensive locally produced timber that meets higher standards. Thus, the US processing industry is not necessarily hurt by all unsustainably produced imports; being able to acquire cheap imports of unprocessed timber is sometimes in its interest. Indeed, between 1990 and 2004 domestic production of industrial roundwood (defined as all timber harvested, excluding fuelwood[15]) decreased by more than 275 million cubic feet, to 15,301 million cubic feet, even though domestic consumption increased by over two billion (2000 million) cubic feet, to 18,737 million cubic feet. This difference was made up by a decrease in exports of over 150 million cubic feet and an increase in imports of almost two and a half billion. Log imports alone increased almost thirtyfold.[16]

The processing industry does have two interests that would seem compatible with an effective international forest agreement. First, it has an interest in protecting itself from imports of wood *products* that come from sources with low environmental standards, particularly because the United States eliminated its own tariffs on wood products unilaterally in the 1980s.[17] However, one way of doing this that has repeatedly been found effective is to raise tariffs against a particular country's products. Thus, for instance, in a later dispute over Canadian exports of lumber to the United States, the US raised tariffs against lumber exports from certain Canadian provinces to over 30 percent at one point in 2001, citing low stumpage fees (the price paid for harvesting standing timber) in Canada as a rationale for accusing the Canadians of dumping unsustainably produced lumber. The US tariffs applied only to semiprocessed lumber and not to raw log imports from the same sources, however, indicating the relatively low importance of the environmental argument in this case.[18]

Second, although the US processing industry's main interest is to protect itself from cheap imports of processed wood products, given that the United States consumes most of its timber and wood production domestically, it also has an interest in developing overseas markets as these are seen to hold the most promise for growth. Again, though, the focus is on tariffs, in this case tariff reductions in other countries. Tariffs on logs and lumber tend to be low across the board, but many nations impose high tariffs on imported manufactured wood products, putting US wood products exports at a price disadvantage relative to local production. Traditionally the US industry has addressed trade discrimination through trade-related measures focusing on particular commercial products, with specific complaints being handled through the US Department of Commerce or the US Trade Representative's office. The "forest sector" as a whole does not come into these disputes, as there is no one organization that represents the whole sector internationally.[19] In addition, there is also an ongoing "Zero-for-Zero" initiative effort to eliminate all tariffs on forest products through the WTO in order to open more overseas markets to US products, with similar negotiations taking place under regional trade agreements.[20] These would have the effect of leveling the playing field in one important sense.

Thus, an effective binding convention would benefit the powerful processors much less than the less powerful nonindustrial primary forestland owners, in light of the processing industry's own interest in cheap imports in some cases and its ability to protect its interests both domestically and overseas through more traditional trade measures.

One other factor that may have contributed to the US industry's overall lack of interest in an effective international agreement in 1992 was the fear that it would entail costs for the US industry. A convention held some threat of enshrining international standards that might force the industry to change its own production practices or be subject to international oversight of its own activities.[21] This threat was felt despite the fact that the US proposal did not call for any regulation of the US forest industry. This seems ironic in hindsight, given the lack of language in the proposal itself that could be construed as threatening to US industry practices. Although it did not limit itself to tropical forests, the United States' proposal described the problem for the North's forests as one of acid rain, not deforestation itself, as mentioned above.[22] The American proposal said nothing about logging practices in temperate forests, despite the fact that the United States itself was entering a period of great domestic controversy over logging of old-growth forest in its Pacific Northwest region—the "spotted owl controversy," as it came to be called.[23]

The mild tenor of the US proposal led one NGO observer at the negotiations to comment that, "[t]he Americans are keen on a forest convention because they would not have to do much."[24] Notably, during the negotiations themselves the United States did not support a Canadian proposal for language on targets and timetables in national plans for sustainable forest management.[25] Nevertheless, the fear of stringent international standards has also been stated as one reason that the US industry organization decided to oppose actively further negotiations toward a global forest convention after the Clinton/Gore administration came to power in the United States.[26]

There was another factor working against any successful completion of a binding global forest treaty: US officials already knew that Malaysia, as well as perhaps other large forested countries such as Brazil, would present obstacles to such an endeavor, given the history of international efforts to curb tropical deforestation in the 1980s.[27] Given the lack of interest in a treaty from these key actors as well as the US forest industry itself, why propose one at all?

More than one US official has expressed cynicism about the reasons for it.[28] Probably the main attraction of a proposal for a forest convention for the United States had to do with its relationship to negotiations on climate change. Numerous observers, authors, and official representatives of other countries speculated in the early 1990s that the United States' position on forests emanated from a desire to divert attention from the climate change issue in the run-up to UNCED.[29] But it was not only a diversionary tactic. Because the US proposal carried few costs for the United States, it was thought that focusing international attention on forests as carbon sinks was also a virtually costless way to be seen to *address* the issue of climate change, which otherwise threatened costly changes in the American lifestyle.

In 1990, as the climate change negotiations were getting underway, the EC (European Community) and the G-77 appeared to be on the verge of making a deal on "tit for tat" protocols within the climate change regime. The EC was proposing a "forest protocol" to the Framework Convention on Climate Change (UNFCCC) then under discussion, which would go to forests as carbon dioxide sinks within the climate change context.[30] The G-77 was proposing an "energy protocol" in order to shift attention back to the North as the source of carbon emissions, the primary suspected causal factor in climate change. Although the EC was prepared to make a *quid pro quo* deal with the developing country proponents of the energy protocol, the United States was adamantly opposed. Any proposed regulation of greenhouse gas sources held an immediate threat of enormous costs for the United States in particular as the world's largest user of energy and producer of carbon emissions.[31] Because the threatened costs were so high for the United States, the US government was prepared to forestall any such protocol deal even if it meant negotiating an entirely new international instrument. Thus the idea was hatched to nip the protocol deal in the bud by superseding it with a full forest convention:

> The US felt it absolutely had to kill the energy protocol and the only way to do that was to also kill the call for a forest protocol, which was by our allies. So how do you do that? The way to do that is to say, "Forests are too important. They provide a wide range of goods and services. And therefore they should be negotiated in a free-standing convention of their own." This kills the forest protocol, which in turn kills the energy protocol.[32]

The juxtaposition of these perceived benefits with very little perceived cost resulted in a US proposal for a global forest convention, produced in July of

1990. Referring explicitly to the highly successful 1985 Vienna Convention for the Protection of the Ozone Layer as a model, the US proposal called for negotiation of a convention to be ready for signing at UNCED and included suggested elements for such a convention. First, on research and monitoring, the proposal called for language on research into forest management practices, cost-effective reforestation techniques, and sustainable yield strategies, for a worldwide network to monitor the health and deforestation rates of the world's forests, and for an inventory of forest resources for new products and uses. It also recommended language on technical training and assistance. The proposal envisaged use of the convention to develop strategies for reforestation and rehabilitation of forests, and to "lay the groundwork for bilateral and multilateral agreements" on air pollution. With regard to financial assistance, the proposal suggested that the convention could address the need for review of bilateral and multilateral assistance programs to explore ways to promote sound forestry practices and ensure that assistance programs do not adversely affect forests. The proposal also called for the convention to explore devices such as debt-for-nature swaps and local currency environmental trust fund programs. Finally, it pointed to the need to change subsidies and market distortions that encourage deforestation or discourage afforestation, such as subsidies that encourage conversion of marginal lands into agricultural lands.[33]

Other Countries

Once the United States had decided to promote a forest convention, it was faced with the task of getting other countries on board:

> Our basic idea was to have a treaty among 35 to 40 or so key forest countries, North and South (Bill Reilly[34] was the father of this notion) and not involve all the unimportant riff raff. If you want to conserve forests, the rest don't matter. In the end, it was impossible to work outside the UN, which is everyone.[35]

Initially the lines between support and opposition were somewhat unclear, as a number of countries weighed competing interests. However, positions eventually coalesced along sharp North-South lines.

Bringing Developed Countries on Board

The United States went first to the other G-7 countries in order to get help from those countries in putting it on the international table. The forest convention was officially proposed at the Houston G-7 Summit in 1990. One author has noted that the proposal for the convention was juxtaposed with an

American statement of support for a German-led "Pilot Programme for Brazil" (PPB) at the summit:

> At the press conference after the Houston summit, Bush declared that he proposed to start negotiations on a global forest convention and that Kohl had been a strong proponent of the PPB.[36]

In fact, there were explicit links between the PPB proposal, the negotiations on climate change, and the proposal for a world forest convention. Due to the perceived stakes involved, President Bush fought for the forest proposal at the summit. Even so, it was a hard struggle. All the other G-7 countries rejected the idea, until Bush met privately with German Chancellor Helmut Kohl:

> Chancellor Kohl had a pet project, with [Brazilian President] Collor . . . a G-7 pilot program to conserve the Brazilian rainforest [the Pilot Programme for Brazil]. Kohl wanted the G-7 to launch it with Brazil. . . . Bush and Kohl went in the back room and Bush said he'd support it if Kohl would support a "global forest agreement" in the [G-7] communiqué.[37]

Germany thus came on board, and with that support the United States was able to get an agreement from the other G-7 leaders. The joint communiqué issued at the end of the summit announced that they were

> ready to begin negotiations, in the appropriate fora, as expeditiously as possible on a global forest convention or agreement, which is needed to curb deforestation, protect biodiversity, stimulate positive forestry actions, and address threats to the world's forests. (Houston Economic Declaration, paragraph 67)

This terminology was carefully chosen, as not all the G-7 leaders were completely comfortable with the idea of a stand-alone convention by the end of that summit. First, Japan was never fully convinced of the desirability of a forest convention. Japan had been under pressure from environmentalists during the 1980s because it was the world's largest importer of tropical timber. As part of its effort to demonstrate environmental sensitivity, it had been a key player in the negotiations of the ITTA and hosts the headquarters of the International Tropical Timber Organization, the intergovernmental organization created under the ITTA that is mandated to address problems concerning the international tropical timber trade. In addition, the United

Kingdom, Germany, France, and Italy still really wanted a forest protocol under the prospective climate change convention, so the idea of twin protocols was not abandoned elsewhere in the communiqué: "Work on appropriate implementing protocols should be undertaken as expeditiously as possible and should consider all sources and sinks" (paragraph 63). The use of the words "convention *or* agreement" in the announcement (paragraph 67) also left open the idea of a protocol under the climate change convention. The idea of a forest protocol to the climate change convention was still promoted even after the G-7 Summit, such as in the conference statement from the Second World Climate Conference (the forerunner to the International Negotiation Committee for climate change) in Geneva in October 1990.

The United States, however, focused on the words "appropriate fora" as entailing a new free-standing agreement and ran with that interpretation. After getting the G-7 endorsement, however weak, the United States proceeded to attempt to win over other countries through *demarches* and diplomatic efforts to persuade them of the merits of the idea. The central question that needed answering was, "What are the merits for a free-standing convention on forests as opposed to a protocol under a climate change convention?" The argument that was developed focused on the need for "a comprehensive, holistic approach that would take into account the multiple values and benefits, economically and ecologically that forests provide."[38] Talking points put together at the time for US negotiators stress the distinction between a convention and a protocol:

> A separate Forestry Convention[39] would address important issues concerning worldwide forestry. These issues are broader and more fundamental than what might be included under a separate protocol relating to global climate change. Such issues warrant the attention of a separate convention strategy—as the importance of protecting forests and advancing reforestation extend far beyond the impacts a convention would have on potential climate change. We should move forward on forestry because it makes sense in its own right.[40]

Most countries considered the new idea of a free-standing forest convention an unwanted distraction from the myriad issues already under negotiation for UNCED. The other OECD countries were eventually convinced of the merits of the idea, however, especially as the G-7 were on record as supporting the proposal. The idea had the support of agricultural ministries in many countries, who manage forests in most countries and who may have thought that a forest convention would lead to greater funding or other support for their work and who then lobbied their own governments. Acceptance of the proposal was also helped by the fact that a traditionally anti-regulation state, the United States, was leading the arguments in favor of international regulation of forests, trying to win the favor of countries already more predisposed toward regulation in many cases:

> We started with the G-7 to get the idea on the table by more than the US. It was a good place to start since we manipulated the outcome we wanted. We then went out with *demarches* to everyone, but in the end, we just got the regulation lovers—the EU can never resist a treaty.[41]

Once the forest and climate change issues had thus been successfully delinked, one primary benefit of proposing a global forest convention had now been achieved. This may constitute the first signal that the United States ultimately would not be willing to pay the full costs that achieving an effective convention would entail.

The G-77 as Anti Coalition

While the OECD countries were brought on board relatively easily, the G-77 were another story. Among developing countries there were several early reactions. Some developing countries in Africa and Latin America favored a global forest convention at first, although with no firm conception of what it would entail. India suspected, rightly, that the proposal had something to do with reluctance on the part of developed countries to take firm action on greenhouse gas emissions and their desire instead to use tropical forests as carbon sinks. Brazil, with the largest extent of tropical forested area in the world, raised objections but was generally more noncommittal than might have been expected from a country that was home to 30 percent of the world's intact tropical forest. As host of UNCED, Brazil tried to project a more environmentally sensitive and conciliatory image during the negotiations than it might otherwise have done, out of concern for the success of the conference (see Porter and Brown 1996, 102; Kolk 1996b, 157–158; Taib 1997, 78).

According to Taib (1997, 45), it was Malaysia, the world's largest producer of tropical timber and the country singled out by NGOs as the worst offender in the destruction of tropical forests, that was most concerned about the threat a convention posed to what it considered its sovereign right to exploit its forests, particularly for international trade.[42] In cost-benefit terms, Malaysia feared that the costs of actions required under a forest convention would not only be excessive in their own right but would also cut drastically into their export earnings and longer-term development plans based upon their predicted timber export revenue.

Timber revenues amounted to 12 percent of Malaysia's total exports and the value of the remaining rainforest area was estimated at almost $70 billion in 1992. Rather than providing an incentive to maintain this renewable resource in order to maintain income, the economic value of the resource was officially considered to be less than the value of the investment in development projects that could be funded from export revenues from "mining" timber (see Kolk 1996b, 155). Malaysia was thus in a unique position in having an explicit official policy of deforestation for development. According to Porter and Brown (1996, 126), Malaysia "was determined to become a fully developed country by the year 2020 using export earnings from timber and other exports crops grown on land converted from forests" under their Objective 2020 program. It was predicted that Malaysia would deplete its entire forest resource by the year 2000.[43]

Another catalyst for Malaysia's extreme position on sovereignty was the backlash Malaysia had felt from the international press and many countries for logging in indigenous lands in the Malaysian state of Sarawak, which in fact hurt its market share. In response, Malaysia invited the ITTO (whose Executive Director was Malaysian) to send an investigatory mission to Sarawak, to assess the status of forest management there and to recommend measures to strengthen it. The mission was mounted in 1989/1990, thus having just taken place as the Forest Principles negotiations started up, and the Malaysians were not pleased that it had found much to criticize (Gale 1998). This whole experience of international scrutiny and criticism of its forest practices did much to sensitize Malaysia regarding sovereignty.

Sovereignty was not just an issue for Malaysia. Most of the countries of the G-77 have taken strong positions regarding sovereignty over the years, particularly over their natural resources, due to the struggles they endured to overcome their exploitation by foreign powers during colonial times (Mickelson 1996, 242–249). Another factor in numerous countries' assertions of sovereignty is that in numerous developing countries, governments simply do not have the administrative control with which to implement international commitments. In these cases the insistence on absolute sovereignty—displayed by countries such as Brazil and Indonesia as well as Malaysia—"is a natural defence against such difficulties" (Grubb et al. 1993, 36). For all these reasons, a preoccupation with sovereignty dominated the G-77 in the forest negotiations, turning this caucus into a hostile anti coalition with Malaysia in the lead.

This issue of sovereignty became a major area of debate in the forest negotiations themselves, which took on a decidedly North-South tone. The G-77 argued that Southern sovereignty over forest resources was under attack from Northern countries intent on making forests into a global issue in order to get control of them. Given that the largest remaining stands of any type of forest in the world were (and are) located in developing countries, there is a ring of truth to this.[44] The South, led by Malaysia, saw this as a "message [by developed countries] that developing countries did not know how to manage their forest resources and therefore they would have to take the lead," as reported by a Malaysian delegate to UNCED (Taib 1997, 79). Either that or, as Malaysia viewed it, industrialized countries were using the issue of conserving tropical forests to prevent rapidly industrializing countries from achieving developed country status (Porter and Brown 1996, 126, fn 50). Malaysian Ambassador Razali Ismail accused developed countries of "imped[ing] the full utilization of our resources and put[ting] them at the disposition of the transnationals."[45]

Meanwhile, whether masking ulterior motives or not, the North used the rationale of forests' global values in advocating a global forest instrument. Both the United States and the FAO made much of forests' carbon sink services and their unique biodiversity, especially in the tropics. It was these global values, as well as the global aesthetic and existence values of the great forests of the world, that first caused the wave of public outcry against tropical deforestation, ultimately leading to the issue's placement on the international agenda. Thus the conceptualization of forests as the "common heritage of mankind", to acknowledge the value placed on forests by people who receive no other benefit from forests. For good or ill, this conceptualization of forests implied that people in the North were entitled to some of the values inherent in forests in the South and thus should have some say in their use. The South countered that forests were a local, national, and developmental issue; as such, they were a national problem and should be exploited in line with national policy. India's Minister for Environment and Forests, Kamal Nath, frequently compared forests to oil resources throughout the negotiations, arguing that if forests should be a globalized commodity because of their global value in absorbing carbon emissions, then crude oil, the most crucial resource in industrialized society, should be similarly globalized.[46]

There was another facet of this dispute for the South: the fact that temperate countries appeared to be taking the lead on tropical forest management was somewhat ironic, given that temperate forest countries have by and large lost a far greater proportion of their original forest over time than the tropical forest countries have. Even as late as during the run-up to Rio, the Bush administration was working to open logging activities on four million acres of government lands in the Pacific Northwest region that contain much of the last old-growth forest within the United States (Panjabi 1997, 108) and was coming under increasing criticism from American environmentalists. As one official puts it, the Pacific Northwest controversy seemed to demonstrate that the most sophisticated and wealthiest democratic country in the world was unable to resolve its own conflicting demands for limited resources, even as it was advocating international commitments for developing tropical forested countries to address very similar issues.[47]

Numerous commentators have stressed the issue of sovereignty as being decisive in the failure of the forest negotiations. Humphreys (1993), for instance, emphasizes the "national resource-global heritage divide", to label the polarization of the international community over whether states or the wider international community have the more legitimate stake in forest conservation. Porter and Brown call the desire to avoid any restrictions on exploitation of national forest resources the top priority of timber-exporting countries such as Malaysia and Indonesia, and argue that it prevented the South from using their forests as leverage to win concessions from the North because they were unable to make concessions to Northern concerns on forests in return (1996, 117). One US official goes so far now as to agree with Malaysia's statements at the time, taking the position that the sovereignty issue led to the failure of the forest negotiations:

> Forests are not naturally globalized, no matter how much you want to talk about them in the global context. Forests are tangible, local, you know where they are, often who they belong to; they don't move around except in international trade. And the most global aspect of forests are biodiversity [which was already being addressed in the negotiations for a biodiversity convention], the carbon sequestration values [already being addressed in negotiations on climate change], or trade [addressed within the GATT, the ITTA, and other international trade agreements]. What's left after that is not global.[48]

There thus existed an asymmetric deadlock at the beginning of negotiations for a global forest convention. Given the United States' interest, the question then became whether the United States would be able or willing to manipulate the preferences of the anti coalition in order to complete an effective, binding agreement.

The Negotiation of the Forest Principles

The actual negotiations for a global forest instrument took place over five meetings: four preparatory committee (PrepCom) meetings leading up to UNCED and the UNCED itself (see Timelines, Appendix 5.1). From the beginning the negotiations on forests were on shaky ground. One major difficulty of the negotiation process was that it was tied to the general

PrepCom process for UNCED as a whole. The main work of the PrepComs was the negotiation of the vast international action plan, Agenda 21, along with the general statement of principles on environment and development that became known as the Rio Declaration. Each PrepCom was divided into three working groups: Working Group I covered the atmosphere and land resources, and included forests under its purview; Working Group II covered oceans, seas, and coastal areas; and Working Group III (created at PrepCom 2) dealt with legal and institutional matters. Given the huge number of issue areas to address, negotiators could not give forests the attention warranted by the perception of deforestation as one of the major environmental problems of the day. In addition, the fate of the forests became inextricably linked to other issues in the broader Agenda 21 context, particularly on finances and technology transfer.

PrepCom 1

During the first round of negotiations for UNCED it became apparent that there was still a gaping lack of consensus on whether there should be a forest instrument. Among OECD countries the question was whether the instrument should be a protocol or a convention, with only the United States and Canada coming out strongly in favor of a convention. As for the G-77, most developing countries were unprepared for what they considered an onslaught by the developed countries on the subject of forests (Taib 1997, 78), and were uncertain how to react. As a result, Malaysia dominated the developing countries' position at PrepCom 1 and from that point on was able to mobilize and lead a developing country coalition of veto states that vehemently opposed the idea of negotiating a binding forest convention. As the Malaysian representative warned the G-77 at a special G-77 meeting on "Combating Deforestation" during PrepCom 1,

> [w]e need total and full support to strengthen your hand at the meeting with those outside our group. Our position should be to first obtain relevant data and encourage dialogue through existing mechanisms. Any legally binding instruments such as the proposed forest convention should not be concluded in haste without taking into consideration their implications on developing countries.[49]

The only agreement reached at PrepCom 1 was that the UNCED secretariat should prepare a comprehensive report on the roles and functions of forests (Decision 1/14); this was done in December 1990 (Humphreys 1996a, 91).

After PrepCom 1, developing states were further antagonized by what they saw as manipulation of the FAO by the developed, pro-convention states (Taib 1997, 80). Not only had the FAO been the first body to endorse a stand-alone forest convention, even before the United States put forward its proposal,[50] it now took the initiative to draft a document on "Possible Main Elements of an Instrument for the Conservation and Development of the World's Forests"[51] and tried to push this draft at two FAO meetings in September and November 1990. This despite having received no mandate to do this from either the states that formed its own decision-making body on forests (the FAO Committee on Forestry, or COFO) or from within the UNCED process. This document, although deliberately entitled in a vague manner, actually gave indications that it was intended as the basis of a convention, including referring to itself as a "Forest Convention" in a footnote (UNFAO 1990, draft Article V.8, sub-paragraph c, fn 4).

PrepCom 2

The biggest point of contention at PrepCom 2 was the FAO draft and its influence. In effect the first full draft text of a global forest convention ever produced, the paper influenced a draft document presented by the UNCED secretariat itself at PrepCom 2. The UNCED draft was totally rejected by Malaysia's Ambassador to the FAO, Ting Wen-Lian, who headed the Malaysian team negotiating forest issues.[52] An *ad hoc* subgroup was therefore formed at PrepCom 2, to submit agreed proposals and recommendations to the Working Group. Here Malaysia, although rejecting the idea of a convention outright, presented a list of "points of concern" (Humphreys 1996a, 91), or demands. This included the need for technology transfer and additional financial resources as "compensation for opportunity cost foregone"[53] to the extent that any convention would commit developing countries to halting or substantially slowing deforestation by reducing timber extraction, agricultural development projects, or conversion of forests for subsistence farming (Taib 1997, 80). Another "condition" for cooperation on forests was a commitment by developed countries, including the United States, to reduce their energy consumption and to provide funding and technology transfer for developing countries to control their greenhouse gas emissions (Taib 1997, 80–81).

The fact that Malaysia itself was willing to provide a list of demands at PrepCom 2 indicates a possible willingness to use its forests as leverage to claim value, knowing that the United States in particular had made it clear that it was eager to have a convention on the world's forests.[54] Indeed, in an interview for the *Earth Summit Update*, the head of the Malaysian delegation, Razali Ismail, said that studies on "the importance of forests in economic and social development and quantification of the economic values of forests," which had been called for in a PrepCom 2 decision, "could be the basis of eventual negotiations on a forest agreement." By "agreement" it is clear that the Ambassador was not talking about the Forest Principles themselves, because at PrepCom 2 the G-77 had already agreed to examine at PrepCom 3 a "non-legally binding authoritative statement of principles," even while "express[ing] strong reservations about trying to reach any forest *agreement*"[55] in time for UNCED.[56] This may have indicated that the value of effective agreement to the anti side could have been raised by linking acceptance of some of the South's demands to the pro coalition's desire for agreement. Humphreys reaches the same conclusion:

> The South reiterated . . . claims [for debt relief and technology transfer from North to South], using forests as a bargaining chip in an attempt to reach a trade-off. Hence the Malaysian intervention in PrepCom 2 . . . can . . . be seen as an attempt to introduce forests as a bargaining chip. (Humphreys 1996a, 96)

The United States and other industrialized countries responded, however, by indicating "their unwillingness to provide tropical forest countries with additional resources, arguing that the sustainable management of forests was in the developing countries' own economic interests" (Taib 1997, 81).[57]

The *ad hoc* Subgroup submitted a document to Working Group I that made no agreed recommendations but instead issued a "Draft Synoptic List" with ten options for a global forest instrument, most mutually exclusive.[58] This lack of consensus led to Decision 2/13 of PrepCom 2, which called on PrepCom 3 to "examine all steps towards and options (including at a minimum, taking into account the special situation and needs of developing countries, a nonlegally binding authoritative statement of principles) for a global consensus on the management, conservation and development of all types of forests."[59] It has been reported that this language, obviously reflecting strong disagreement and itself heavily negotiated, took up most of the time of the PrepCom (US official, personal communication, August 2001). It reflected a compromise between those who wanted a convention and those who did not, but it showed a swing away from a convention and towards a nonbinding instrument. The United States, however, did not concede on this point.

Meanwhile, the conflicting positions that had been heard at PrepCom 2 were echoed in two documents issued between PrepCom 2 and PrepCom 3. In Beijing in June 1991, ministers from 41 G-77 countries and China formulated a "Beijing Declaration" in order to build a coherent and unified G-77/China position for the UNCED negotiations. On forests, areas of possible linkage are made clear in the Declaration's statements that any multilateral measures to conserve forests must also "enhance the economic, social and environmental potentials of forests" and ensure better market access for value-added timber products, and in its call for industrial countries to increase their own forest cover.[60]

Although not formally intended as a response to the Beijing Declaration, the second document issued during the intersessional period can be seen as playing that role. This was an official proposal from the United States for "Principles for a Global Forest Convention/Agreement," which it sent to the G-7 countries and more than 20 tropical forest countries.[61] It called for "stewardship," as in "a responsibility to engage in cooperative stewardship to improve global environmental quality for mutual benefit,"[62] which the South found unacceptable given its position on sovereignty. It also called for reliance on existing bilateral and multilateral assistance agencies rather than any "global funding mechanism" for programs to increase forest cover, as preferred by the G-77.[63] It also reflected strong domestic pressures to continue cutting down primary forests in the Pacific Northwest. The proposal received much criticism from American environmental NGOs as well as from tropical forest countries and had little, if any, impact on the remaining negotiations.

PrepCom 3

At PrepCom 2 the UNCED Secretariat was given the task of analyzing and addressing the items outlined in the Draft Synoptic List. It formulated a document in April 1991, "Guiding Principles for a Consensus on Forests" (PC/65),[64] to be used as the basis for negotiation at PrepCom 3. In places this document drew directly from the FAO draft convention, which once again did not sit well with the G-77. The G-77 therefore presented its own proposal at PrepCom 3 (L.22),[65] which took its title verbatim from Decision 2/13. The G-77 won on this point and at PrepCom 3 the Working Group agreed to negotiate toward the goal of a set of non-legally binding forest principles, in other words, the minimum established at PrepCom 2.

The G-77 draft also introduced claims for technological and financial transfers, the financial resources to be given as "compensation for opportunity cost foregone." The draft used very similar wording to statements made at PrepCom 3 by Malaysia and India, suggesting their continuing leadership of the G-77 in the forest debate.[66] The G-77, in a new partnership with China, issued a joint proposal on financial resources, to be "compensatory in nature". This, along with statements made at PrepCom 3, other concepts being introduced such as a "partnership in additionality,"[67] and a proposal by China for a "Green Fund"[68] provide further evidence that sovereignty over forests was not the exclusive concern of the South but was being used as "environmental leverage".[69]

The United States, supported by Australia, the EC, and Japan, countered by promoting the Global Environment Facility (GEF) as the sole mechanism for funding conservation of the world's forests and conditioning GEF funding on donor approval of projects and evidence of "good governance" in the recipient country. The GEF, mentioned in chapter 3, was created by the World Bank in November 1990 with a three-year experimental grant of $1.3 billion; forest conservation and management would only be funded through the GEF insofar as it related to the global benefits of biodiversity preservation and climate change mitigation, two of the four GEF windows. It garnered much criticism from developing countries, however. Being a product of the World Bank it was dominated by that institution and its practices reflected that influence: It lacked transparency, its executive officer was a Bank official, and voting was weighted in favor of the donors. The largest sums it had allocated had been channeled through the World Bank, even though UNDP and UNEP were brought in as co-implementing agencies to appease the South. GEF projects also held the risk of being influenced by World Bank "conditionality", whereby financing might be conditioned on the target country's meeting certain "sometimes draconian" conditions.[70]

Compared to the $125 billion in external assistance that the UNCED secretariat calculated would be necessary to meet the obligations of Agenda 21 (Kolk 1996a), the GEF's potential, in any case, was miniscule at best. More importantly, however, given its ties to the World Bank, developing countries viewed the GEF as a way to "thrust the priorities of industrialized countries down the throats of the developing countries."[71] The South used their desire for noninterference in their sovereign affairs as a rationale for rejecting this idea, and positions began to harden.

In the end, documents PC/65 and L.22 were merged by the UNCED secretariat, along with comments made at PrepCom 3, and the synthesis document formed the basis of the draft Statement of Forest Principles for subsequent negotiations (Humphreys 1996a, 97; Taib 1997, 81). It has been noted, however, that in reality the UNCED secretariat used the G-77 draft as the overall guiding framework. According to one US official involved, "The forest principles were based on a G-77 draft text [of nonbinding principles] drafted primarily by Malaysia, India and Brazil. The donor countries had, as usual, to work off that. So we did."[72] Humphreys confirms that the consolidated draft "overwhelmingly reflected G-77 concerns," ascribing this to the G-77's numerical superiority (Humphreys 1996a, 97–99).

Developed countries, meanwhile, shifted their sights from negotiating a convention for Rio to negotiating basic principles upon which a subsequent convention could be based, as reflected in bracketed language of the consolidated draft:

> The UNCED process is the most appropriate forum for conclusive decisions pertaining to global consensus on forests [which should form the basis for [any subsequent preparations and adoption of a legal instrument on forests] [all other negotiations involving forests]].[73]

Nevertheless, some members of the American delegation, despite acknowledging at PrepCom 3 that a binding agreement would not be achieved by UNCED, continued to hold out hopes that a legally binding convention on forests could be accomplished as early as the end of 1992.[74]

Interestingly, references to forests as the "heritage of mankind" began to be made in G-77 contexts after PrepCom 3. For instance, at a "G-15" summit meeting of the fifteen largest developing country powers, held in Caracas in November 1991, Venezuelan President Carlos Andres Perez stated that

> [i]nordinate responsibility cannot be placed on efforts to preserve the environment, whilst the South has limited access to, or is excluded from, technologies we need to back our national development efforts and fight poverty. If the tropical forests are the heritage of mankind, science and technology should be also.[75]

Reference was made to both forests and biodiversity as a common heritage of mankind in the "African Common Position on Environment and Development" that was adopted at the Second Regional African Ministerial Conference on Environment and Development in Abidjan in November 1991.[76] Humphreys refers to these references as indications that at least some G-77 actors were willing to use forest sovereignty as a bargaining counter in exchange for a *quid pro quo* from the North, even though the common G-77 position was that sovereignty could on no account be compromised.

All of these developments and statements by developing country groups suggest that opposition to a forest convention was not monolithic and bolster the idea that there may have been some bargaining room on the statement of forest principles at PrepCom 4; however, meaningful bargaining did not take place.

PrepCom 4

The draft text of Forest Principles arrived at PrepCom 4 with 116 sets of brackets around disputed text. Unlike the Framework Convention on

Climate Change and the Convention on Biological Diversity, it was forwarded to UNCED with 73 brackets remaining, including a set around the crucial (to the United States) call for future negotiation of a world forest convention. Simultaneously, an unexpected proposal on a future forest convention was again being made by the United States, not within the Forest Principles negotiations themselves but within concurrent work being done on the Agenda 21 chapter on deforestation.[77] One US delegate to PrepCom 4 recalls:

> [Agenda 21] was negotiated by a guy from the US Forest Service . . . probably not very well. We just let him do Agenda 21, because for us that was not *the* negotiation on forests. The negotiations on forests were the Forest Principles. That was the policy/political negotiation. We've actually never paid any attention to Agenda 21. [The negotiator] was in line with the US position on a convention but it may not have been timely to mention it when he did.[78]

This threw the Forest Principles negotiations into chaos as well, provoking a stormy response by Ambassador Ting. Again, though, her response appears to leave room for negotiation on the issue of sovereignty:

> We wish to underline the supremacy of our sovereignty over our forests. We are certainly not holding them in custody for those who have destroyed their own forests, and now try to claim ours as part of the heritage of mankind. . . . Our message is clear; we are prepared to play our part in the great environmental effort. We are prepared to sustainably use our sovereign forest resources. However, we require financial resources and technology to carry out our environmental obligations.[79]

This was once again met with a decided lack of response on the part of the United States. As one US negotiator at PrepCom 4 notes:

> For the US the negotiation of the Forest Principles was extremely difficult. Our position was this: no money and no technology transfer, [but developing countries must] conserve tropical forests. [Meanwhile], we were at the height of the Pacific Northwest [situation]. . . . [W]hat an incredible issue it was! [I]t tore the country apart, and it influenced negatively our ability to negotiate so much, and at the same time we're saying "No, we're not going to agree to new foreign money." Everybody knew about [the Pacific Northwest conflict] and was saying, "You have the greatest country in the world and it cannot resolve a local community issue over conflicting demands for a limited resource, and you want *us* to do this?" It was horrible!

> [Especially with the G-77 saying] that we had destroyed our forests, it was all our fault, and they should be compensated.[80]

There was one further meeting of G-77 countries between PrepCom 4 and UNCED itself, in Kuala Lumpur, in April 1992. Here the Malaysian prime minister again seemed to give an opening: "If it is in the interests of the rich that we do not cut down our trees then they must compensate us for the loss of income."[81] Malaysia even proposed a "Greening of the World Initiative" as a mechanism for this, calling upon the global community to target at least 30 percent of the earth's terrestrial area to be greened by the year 2000. The percentage of land under forest cover worldwide already constituted 27.6 percent, so the Malaysians asserted that a target of 30 percent was realistic. Those countries that did not have suitable land area should instead contribute adequate funds to developing countries with available land, through the previously proposed global Green Fund.

The proposal was clever in allowing Malaysia to claim to good environmental stewardship: Malaysia still had 56.2 percent of its land under forest cover and pledged that it would only deforest 6.2 percent, leaving 50 percent permanently forested. This was 20 percent more than the initiative itself would require and vastly more than the forest cover of many Northern countries. Malaysia could thus get in a dig at those countries that had already largely destroyed their own temperate forests for national economic development and were now seen as trying to lecture the developing countries (Taib 1997, 56–58), while still actually continuing its own deforestation. Several Northern commentators have noted the fact that the Initiative talked in terms of targets, mechanisms and monitoring requirements. This contrasted strongly with the United States' refusal to countenance such intrusions on its own sovereignty and also with the notion that Malaysia's intense opposition to a forest convention was based on a claim of absolute sovereignty.[82]

The final declaration of the meeting reaffirmed the previously stated positions of the G-77 and China, alluded to the Green Fund and the goals of the "Greening of the World Initiative" proposal, and called on developed countries to avoid unilateral measures such as bans and restrictions on the international trade of forest and forest-related products from the developing countries.[83]

The UNCED Endgame

At UNCED itself, forests were discussed in one of eight contact groups set up for all issue areas in which substantial negotiation was still necessary. Even though the South had scuttled the proposal for a binding agreement, and despite the United States' assertions that the South could not be swayed from its obsession with sovereignty, forests in fact remained a strong bargaining chip for the developing countries. First, the North wanted some kind of forest agreement to come out of UNCED, even if nonbinding (although for the South, too, there was an interest in not being seen to be the cause of failure to conclude the agreement, especially as the text bore so much resemblance to the G-77 draft). Second, after its failure to get agreement on negotiating a forest convention for UNCED, and the failure of its continual efforts to "get the Forest Principles kind of converted into a convention as they were being negotiated,"[84] the North wanted to insert language into the Forest Principles to call for negotiations and adoption of a future legal instrument on forests. Finally, the South had been successful in keeping forest negotiations within the UNCED, which opened the door to a range of possible linkages. The South used this leverage over forests to try to press its demands in the larger UNCED context, particularly regarding Agenda 21 language on financial resources.[85]

Even some Americans recognized that forests were being used as leverage, held hostage to value-claiming issue linkage by the South. As the US chief negotiator at UNCED observed at the time, "Some countries are reluctant to take concrete steps to preserve their forests. They are trying to get the money before agreeing to do anything."[86] This observation does not jibe with the (later) US assertions that the sovereignty issue was an insurmountable obstacle to agreement.

In order to attempt to address Southern demands, the United States took the initiative on two relatively low-cost linkages in Rio. First, the United States took a small step toward additionality with its proposal for a "Forests for the Future" Fund. According to a US official, this proposal was the result of recognition within the Bush administration in the run-up to Rio that "we [the United States] had this untenable position of 'conserve tropical forests, no money, no technology transfer, and don't mention old-growth [forest because of the Pacific Northwest controversy]." The US proposed a $150 million (125 percent) increase in bilateral assistance to forests through new appropriations, and challenged other donor governments to double their own forest-related aid.[87] According to one US official, the proposal did help achieve the conclusion of the Forest Principles in Rio and made the United States look less bad, although the United States was disappointed at the time that it did not lead to an agreement on negotiating a treaty later. According to William Reilly, the initiative was a way of inducing the developing countries' acceptance of the proposed Statement of Forest Principles, which he hoped would lead to a forest convention.[88] Developing countries themselves considered it an attempt to persuade them to accept both the importance of forests as carbon sinks and the need for a (future) forest convention (Kolk 1996a, 12).[89]

That the proposal was not more effective might be explained by two factors. First, the fact that it was really less than a drop in the bucket compared to the almost $6 billion ($5.7 billion) per year that Agenda 21 was seeking from developed countries specifically to help developing countries combat deforestation. (Agenda 21 called for a far greater contribution from developing countries themselves, at over $25 billion per year.) Even if the Forests for the Future challenge had been fully met by all the other OECD states, as President Bush envisioned, it would have increased worldwide government spending on forest conservation only from $1.35 billion in 1990 to $2.7 billion (Lane 1992, 79). Second, it was not clear at the time whether the initiative represented new funds for forest conservation projects or merely a reallocation of existing funds (Kolk 1996a, 12–13).[90]

The second US concession pertained to desertification. At PrepCom 4, the African countries, which were generally silent in the forest debate, began to hint of a possible bargain linking African support for a forest convention with US support for a desertification convention.[91] After opposing a desertification convention throughout the negotiations, the United States unexpectedly announced its support for the African position at UNCED itself. According to Porter and Brown, the head of the US delegation, Curtis ("Buff") Bohlen, made the decision on his own with no higher consultations. This coincides with information from another US official:

> He said to me, "These people keep coming up to me and they really want a desertification convention. Is there really a down side for us?" and I said, "Well, we really want a forest convention." Buff said, "Yeah, but we're not going to get that." So you know, he was not really asking my opinion because he was going to do it anyway, but he decided himself, as far as I know, overnight, that he was going to let the Africans get in. . . . This was not in our instructions.[92]

Taib (1997, 84–85) surmises that the United States made this gesture simply because it was tired of being vilified as the conference's "bad guy."[93] He claims that it was the United Kingdom that tried to connect desertification to deforestation in order to get the African countries to break the G-77 ranks and support the proposal for a forest convention.[94] Generally, though, contemporaneous reports drew links between the United States' gesture toward a convention on desertification and its hope of gaining concessions in the Forest Principles talks.[95] The Africans themselves said that although no explicit link was made it was certainly made implicitly. Nevertheless, the president of the G-77, Kofi W. Awoonor of Ghana, maintained solidarity with Malaysia, saying that the question of sovereignty to exploit one's own resources affects all developing countries.[96]

These initiatives did little to shift the anti coalition's position. Southern NGOs at UNCED accused the US of maintaining a position, in effect, that "what I have I keep, what you have I will take," in demanding commitments from the countries of the South on forests while refusing to undertake meaningful commitments on finance and other issues, even in the larger context of Agenda 21.[97] Questions that engendered hot North-South dispute in the Agenda 21 negotiations included whether developing countries' commitments to implement Agenda 21 would depend on provision of financial resources and technology transfer; whether a substantial commitment to financial aid, with firm figures, would be made at UNCED; and whether the World Bank-controlled GEF was the appropriate funnel for environmental assistance.[98]

In the end, the prospect of reaching any agreement, even on the nonbinding Forest Principles themselves, looked at risk. The negotiations were scheduled to conclude by June 11, 1992, before the actual summit of world leaders over the final two days of UNCED. Instead, they continued on behind closed doors at the ministerial level throughout the summit meeting. As of June 12 there were still 14 points under dispute and "a growing sense that the differences between the rich and poor nations on forest use were so profound that the issue might well turn out to be the Summit's most singular failure."[99] The talks finally finished at 3 a.m. on June 13.[100]

Given that the United States had no flexibility on making concessions within the Forest Principles or in Agenda 21 in order to increase the value of agreement for the anti coalition—nor the ability to lower the value of nonagreement through threats of punishment—the only recourse for reaching any conclusion as time ran out on the negotiations was to make the ultimate concession: to drop bracketed text in the Forest Principles calling for the negotiation of a convention. The United States and the pro coalition thus lost their push even for an agreement on the *possibility* of a future global forest convention.[101] The only language even hinting at the possibility was a statement in the Preamble (weaker than operational text), that

> [i]n committing themselves to the prompt implementation of these principles, countries also decide to keep them under assessment for their adequacy with regard to further international cooperation on forest issues.[102]

In dropping language on future negotiations for a convention, the United States won agreement on a Statement of Forest Principles that reflected little or nothing more than the global status quo on global forests. Agreement was reached by deleting some contentious text and weakening other points. While the North gave up a forest convention at UNCED as well as mention of any possible future convention, the South gave up bracketed language on several points. This lost language included a call for the recognition of and compensation for historical forest loss in developing countries; a reference to the past destruction of developed countries' forests and damage caused by acid rain; and a preambular paragraph stating that the adequate management, conservation, and sustainable development of all types of forest requires sustainable production and consumption patterns, particularly in developed countries, and noting the need to ease the debt burden and eradicate poverty. There was little language on financing, as references to the GEF as well as the call for a "Green Fund" were dropped.

The Forest Principles did accommodate the South in some ways, although in much weaker language than desired. Thus, indebtedness and market access for forest products "should be taken into account" in terms of developing countries' efforts to manage their forests sustainably (Principle 9(a)) and transfer of environmentally sound technologies should be "promoted, facilitated and financed . . . as appropriate" (Principle 11).[103] The South even won the statement that new and additional financial resources should be provided to developing countries in order to care for their forest resources (Principle 10). As Heissenbuttel points out, this language "does lend support to the President's 'Forests for the Future Initiative' "—in other words, it did not necessarily represent any cost to the United States above what the United States had already proclaimed willingness to take on.[104] Malaysia and Brazil won two major concepts: the Forest Principles enshrined for the first time that the international trade in forest products should be free and fair—not subject to unilateral actions, tariffs, or bans—and that all types of forest, not just tropical ones, should be subject to the same principles of management.[105] Neither of these principles posed any cost to the United States, as the United States had already eliminated its trade barriers on forest products a decade earlier and the US industry asserted that US forests were already well managed.[106]

Meanwhile, heated negotiations continued on the financial questions that loomed in the Agenda 21 debate right up to the last day of UNCED. In the end, the developing countries gained a total of $2 billion in promises of new funding—compared to the $125 billion needed, as calculated by the UNCED secretariat,[107] and low even when compared with the $10 billion the Brundtland Commission had reckoned as the minimum benchmark by which to measure UNCED as successful. Then, too, it was not clear whether all the contributions announced by donor countries really represented "new and additional resources".[108] On funding mechanisms, the developing countries lost their quest for a new Green Fund under Agenda 21 to replace the GEF, although the GEF was to be reformed in line with developing country criticisms and a need to use "all available funding sources and mechanisms" was also mentioned (Kolk 1996a, 21). A proposal for an "earth increment" for the tenth replenishment of the International Development Association was incorporated into Agenda 21 (Chapter 33.16(a)(i)),[109] although at the December 1992 donor meeting of the World Bank this was rejected. Language on technology transfer was vague. In other areas of potential forest-related concessions, no hard commitments were made in the climate change convention, and no commitments were made on primary forest preservation in the United States.

The Forest Principles: An Assessment

The Statement of Forest Principles has been called a representation of "the lowest common denominator between North and South" (Humphreys 1996a, 102), with " 'progress' . . . almost non-existent" (Kiss and Shelton 2000, 366); and the most "significant setback" of Rio.[110] States still have "the sovereign and inalienable right to utilize, manage and develop their forests . . . , including the conversion of such areas for other uses within the overall socio-economic development plan" (Principle 2(a)). There is, in fact, no mention of a global interest in the world's forests nor any mention of a need to halt global deforestation (Heissenbuttel et al. 1992). The text does not specify policy prescriptions; rather, it stresses policy responses at the national level. It does not indicate how the document is intended to relate to existing international forest processes such as the ITTA/O or the TFAP nor to the new related conventions on biodiversity and climate change. It is also unclear from the document where financing for the activities encouraged by the document would come from.

It has been argued that the Forest Principles are little different from a framework treaty. The main difference would have been language saying "the Parties agree to [do X]" instead of saying that states "should [do X]." In the words of one US official,

> arguably a treaty is more obligatory because it is legally binding. However, in many cases, the process of negotiating a legally binding agreement leads to such a lowest common denominator and generalized obligations that the effect on countries' behavior is considerably less than a voluntary agreement among like-minded countries.[111]

The Forest Principles were not negotiated among like-minded countries in any case, so they still represent a lowest common denominator. There is one significant difference between the Forest Principles and a framework treaty, however, in that a framework treaty holds the promise that more specific commitments will be made later to fulfill the objectives set out, as intended in the framework convention-protocol approach and carried through successfully in the ozone agreements. The Forest Principles held no such potential. Indeed, numerous participants and observers thought at the end of UNCED that there would be no further global dialogue on forests at all.[112]

The notion that US interests are behind the global failure to achieve a binding forest convention, as I argue, must be judged in light of the competing explanation that developing countries' interest in sovereignty prevented thoughtful consideration of the proposal. It is easy at first glance to assume that developing countries "won" at Rio by preventing a treaty from even being put on the agenda for future consideration. Malaysia's Ambassador Ting pronounced the completed Forest Principles "a great victory for the developing nations, especially for Malaysia."[113] As just seen, however, the G-77 lost out on its attempted linkages in both the Forest Principles and in Agenda 21. Did the G-77 really win on sovereignty or did it in fact lose on its demands for financial and technology transfers in the larger context? And how does this relate to the bigger question of the effectiveness of agreement?

Many observers who were at UNCED fault the South for obstructing agreement on forests *and* for not winning themselves more concessions. Porter and Brown's argument that sovereignty was the South's priority certainly fits with Malaysia's claim of victory in the Forest Principles. Panjabi criticizes the South's emphasis on the sovereignty issue that "blinded the developing world to the very important concessions that the North was prepared to make in return for forest conservation by the South" (1997, 144). Sullivan, similarly, argues that "the South was simply not prepared to be told what to do with their forests even if they were offered compensation by the North" (1993, 161). An American participant in the negotiations asserts that "at the time we were disappointed that agreement was not reached at Rio to subsequently negotiate a treaty on the basis of the Forest Principles. But this was not in the cards under any scenario, money or not. The G-77 was not going to commit themselves at that time to negotiating a treaty later, end of story."[114]

The conclusion that the South obstructed agreement must be assessed with caution, however, for several reasons. First, the developing world was generally hostile to the Montreal Protocol as well, as shown above, yet that did not prevent the Montreal Protocol's eventual success. Of course, the reasons for G-77 antipathy to a forest convention were different: unlike the ozone regime, which existed to address a problem of the global atmospheric commons, a global forest convention threatened the sovereignty of forested countries (this was true for all forested countries, not just those with tropical forest). Nevertheless, even acknowledging that the reasons differed, the fact that the developing countries had strong anti-agreement positions on both means that their opposition *alone* cannot account for the failure of the forest negotiations.

Then, too, if Malaysia were seen as the obstacle among G-77 countries, as it was, that did not necessarily doom a forest convention, as there was always a possibility of isolating Malaysia and continuing efforts for a convention with the other G-77 countries. The G-77 was certainly not fully united on this issue. Taib acknowledges that the Declaration of the Second Ministerial Conference of Developing Countries in Kuala Lumpur was only achieved after "a very difficult, exhaustive and strenuous negotiation which continued until the late hours" to develop a common position for the upcoming UNCED, particularly on financial issues (1997, 54–55). Johnson argues that other Southern countries might have been brought over to support a convention, or even to signing a convention in Rio, if OECD countries had "upped their offer and . . . put new money on the table earlier, conditional on the G-77 keeping their side of the bargain" (1993, 7). Malaysia would then have had much less support from other G-77 countries if it had been clear that Malaysian actions were in danger of upsetting the general financial boat. This echoes the words of Angela Harkavy, an environmental consultant who served on the US delegation to PrepCom 4, just before Rio:

> The difficulty in agreeing on a forest convention is perceived to be due to the inflexible positions of the US on climate change, finances, technology and the call for a desertification convention. Some analysts believe that President Bush's presence at the Earth Summit, plus more flexibility towards financial issues, and a kick-off toward a desertification convention could provide a boost for the forest convention. For example, perhaps Brazil could be more flexible toward the forest convention after being honored by the US President's presence in Rio. And some African countries, which were not happy with Malaysian positions on the forest issues, might be supportive of the US's call for a forest convention, if they succeed in winning special attention to their problems, such as in a separate desertification convention. Malaysia, which has been the spokescountry for the G-77 in rejecting forest agreements, would be isolated, and the world could move toward a forest convention.[115]

Of course, as seen above, the United States did meet the call to give support for a separate desertification convention. Not being part of a bigger package, though, it had minimal results, apart from perhaps contributing to the achievement of the Forest Principles themselves.

A second reason for caution in concluding that Malaysia's stance on sovereignty in particular caused the forest negotiations to fail is the question of whether that position really was Malaysia's final word. There was ample evidence that Malaysia was not completely closed to the idea during the UNCED negotiation process, at least not if it had been offered an adequate *quid pro quo*. Grubb et al. assert that the Malaysians did not really object to the concept of a convention but to the *form* of convention proposed by the developed countries, and that their objection was "based not upon a claim of absolute sovereignty but in large part upon differing perceptions of equity,"[116] in other words, a fear that the costs of curbing deforestation would fall too heavily on developing countries.[117] Brazil voiced the same concerns in Rio:

> If it is a convention on international cooperation on the protection of forests, we are all for it. But not if it is a convention that creates obligations which countries do not have the resources or the technology to implement.[118]

Indeed, less than three years after UNCED, Malaysia became an active proponent of a forest convention.[119]

Finally, as Mickelson (1996, 249) argues, even assuming that Malaysia was indeed obsessed with its sovereign rights over its forests, a respect for sovereignty over forest resources need not actually preclude a forest convention. Indeed, the reverse is true: an explicit affirmation of the principle of sovereignty is a prerequisite for developing country participation in a forest convention, and arguably for US acceptance as well. The question is whether sovereignty precludes an acknowledgment that decision-making about the exploitation of forest resources should take into account any interests that go beyond national ones (leaving aside the fact that states in whose territories forests are found also receive a share of the global benefits from forests). A subtle distinction must be made between the legitimate concerns of the international community and the assertion of a more direct international interest in controlling national decision-making over national forest resources.

A competing explanation for the failure to achieve a binding agreement on forests at UNCED must be considered: the lack of true leadership by the United States, defined as the capacity and will to manipulate preferences, in its drive for a forest convention. A number of authors allude to American failure to manipulate values. Porter and Brown, for example, attribute the failures of UNCED, including the failure to gain a convention and the weakness of the Statement of Forest Principles itself, not to the sovereignty stake they earlier cite as behind the South's inability to leverage concessions from the

North, but to US intransigence:

> Only one country—the US—had the economic and political power and the expertise to play the lead role in such a complex and wide ranging set of North-South negotiations, had it chosen to do so. But instead the US played more the role of veto power on those issues. The result was that other OECD countries often were not pushed to move beyond least-common-denominator positions, and developing countries were not offered any real bargains.[120]

Johnson (1993, 7) speculates that the unwillingness of the OECD countries, particularly the United States, "to address their own profligate lifestyles, or even to admit there was a problem," was also a factor in keeping a "global bargain" from being struck at UNCED. In the South's view, overconsumption of resources, such as fossil fuels, by the North was directly linked to forests. The need for substantial concessions by the United States was recognized by at least some in the US Government. Senate Resolution 181, introduced by Senator Al Gore on July 15, 1991, noted the need for US leadership on stemming the destruction of forests worldwide and commitment to meaningful reductions in greenhouse gas emissions, as well as on "innovative financial mechanisms including . . . debt relief to support efforts of developing countries to implement forest protection programs . . ., and initiat[ion of] a national plan to curb deforestation of primary forests."[121]

However, this resolution did not result in material concessions. The North seemed to place an inordinate emphasis on forests as carbon sinks rather than on the problem of carbon emissions produced mostly by inefficient energy use in the North. Thus the South saw the North's focus on a forest convention as a move to pass the burden of the costs of its own profligacy onto the South, restraining the South's economic development (Malaysia's particular concern) in the name of saving trees to absorb carbon.[122] Brazil and India focused on the climate change connection specifically. Therefore, another valuable concession the United States could have made in the name of a forest convention would have been to accept a stronger climate change convention. Instead, the US administration insisted that any mention of proposed targets and timetables for emissions reductions be deleted from the final Framework Convention on Climate Change. This was of course in line with its original motivation for proposing a forest convention in 1990; it won on this point in the FCCC negotiations, but not without causing great resentment among developed and developing countries alike.

As for the alternative possibility of manipulating preferences through threats of action that would lower the value of no agreement relative to the value of agreement, some observers reported threats emanating from the United States in the form of linkages made among the Forest Principles, Agenda 21, and the Rio Declaration. Specifically, Angela Harkavy accused the United States during UNCED of "holding everything the South wants as hostage for the Forest Principles."[123] This implies that the United States was ready to give the South what it wanted *if only* the South would concede to its demands on the Forest Principles. This seems implausible, given what is known about the negotiations—first, that no real possibility of agreement on the G-77's financial demands in Agenda 21 was ever voiced by the United States and, second, that the United States went to Rio with little to offer.[124] In any case, these threats do not match the threats of punishment identified in chapter 2, in that they did not include any threat of lowering Malaysia's position vis-à-vis the status quo.

Panjabi, noting the lack of funding promises in general at UNCED, points out that at that time the developed countries were suffering one of the worst recessions in decades, facing pressure for funding from newly democratizing countries of Eastern Europe[125] and staggering deficits of their own, and making cutbacks in education, health, and welfare programs. As a result, "the basic problem facing all nations that participated in the Earth Summit was the question of funding" (Panjabi 1997, 144). There is actually no credible reason to think that if Malaysia had agreed to go along with the forest convention the pockets of the United States and the North would suddenly have become deeper. Far more plausible is the possibility that it was the developing countries that made the linkage between any acceptance of a forest convention and their demands for concessions, threatening not to sign on to future forest negotiations unless the developed countries met their demands in Agenda 21. This of course did not happen.

Because the game of negotiation requires keeping one's real interests hidden, it may be impossible ever to obtain conclusive evidence on whether the G-77 preferred at that time to maintain the status quo of "absolute sovereignty" over its forest resources, or whether it would have preferred to achieve more in terms of its larger goals. However, one way to investigate competing claims of causality is to use counterfactuals. Assuming, for the moment, that both US intransigence on linkages as well as a Southern obsession with sovereignty, we can assess the relative importance of each by asking what would have happened if one of the factors had existed and the other had not.[126] We state two competing counterfactuals and see whether either leads to a different outcome than what was achieved in reality, attempting to make the argument as plausible as possible. Each counterfactual can be tested by invoking general principles and drawing on knowledge of historical facts relevant to a counterfactual scenario.

The two counterfactuals may be stated as follows:

(1) If it had been the case that, *ceteris paribus*, Malaysia was indifferent to sovereignty claims, there would have been a treaty.[127]
(2) If it had been the case that, *ceteris paribus*, the United States offered leadership in the form of incentives or threats in order to change the G-77's perception of the value of agreement, there would have been a treaty.

Are both true? If not, which? If so, which is of greater relative importance?

Invoking the rationality principle (assuming that the outcome is the result of human choices and that rational humans engage in goal-maximizing behavior), we need to show that one choice is more desirable than other possible choices (Fearon 1991, 182, fn 29). First, would the choice to have a treaty have been more desirable to Malaysia than no treaty if Malaysia were indifferent regarding sovereignty and the United States did little to change its preferences? We know empirically that the incentives the United States did offer had little to do with Malaysia: $150 million to be spread over the globe would barely reach Malaysia if at all, while the United States' support for a desertification treaty, assuming an attempted linkage, was intended to benefit African and other countries with climatic conditions that Malaysia did not share.

The question here is whether Malaysia would ever see the choice of a treaty as desirable without any coercion from the United States. The United States and others reasoned, as mentioned above, that the sustainable management of forests should be seen as economically beneficial to the countries themselves. But as I have earlier pointed out, this depends on discount rate—how much Malaysia and/or other countries value maintaining their forests for the future compared with the present income they can derive from harvesting forest products. We must assume that, even without the label of "sovereignty" as an issue, rational actors in Malaysia and other G-77 states perceived deforestation as more economically beneficial than forest preservation. Thus, the cost of agreement would outweigh its benefits without a shift upward in the value of agreement for them or downward in the value of no agreement. There is no reason to think that Malaysia would ever see a treaty as desirable unless such a valuation shift occurred, either through manipulation by another party or through exogenous events.

One might be tempted to speculate on whether the United States would have given in on financing and other value-claiming issue linkages if Malaysia had been prepared to say yes to a forest convention or to future negotiation of a forest convention. However, this is not really at issue here. The question of US financing will be examined in the other counterfactual scenario. But more importantly, the requirement for US leadership suggests that in order to attain agreement on a convention, or even agreement for future negotiations toward a convention, the motivation would have to come from the United States itself as the proponent, in terms of massaging preferences toward favoring an agreement.[128] Even if Malaysia were neutral as to sovereignty, the idea that Malaysia might agree to a forest convention first, without any sign from the United States of substantive movement toward meeting the G-77's demands or toward otherwise shifting the values of agreement and no agreement for the anti coalition, does not meet the requirement of plausibility.

On the converse counterfactual proposition, if we again invoke the rationality principle and consider which choice seems the most desirable, the question is whether coercion from the United States might ever have overcome Malaysia's fixation on sovereignty. Assuming that the Malaysians were in fact rational actors, we must assume the existence of a point at which the benefits to be gained from an agreement would overcome the cost of maintaining sovereignty. The desire for sovereignty might make an agreement more costly for the pro coalition to achieve,[129] but even then if adequate incentives or threats were applied to make the value of agreement outweigh no agreement for Malaysia, a rational Malaysia should have switched its position. At what point would coercion have worked, and would it have been cost-effective for the United States? Both of these questions can be answered sufficiently by again considering the reality of what the United States did offer. Dropping the call for a forest convention was a much less costly measure for the United States than giving in to the provisions the South demanded, particularly in Agenda 21, in return for such language. In the end, it appears that the prospective benefits to the United States from a world forest convention, or from even a promise of negotiations toward one, were simply not enough to warrant more incentives than those the United States offered at Rio.

Each of these counterfactuals deal with one of the two most concerned countries in the UNCED forest debate. The first posits a different attitude on the part of Malaysia; the second, a different position for the United States. We then ask which counterfactual is more plausible. We find that in the first case, even if Malaysia had not used the sovereignty argument, the possibility of a binding agreement or an agreement on negotiations for a future binding agreement would still have been remote. In the second case, if the United States' position on compromise had been different (assuming the benefits of a treaty to the United States could have outweighed a much costlier treaty), it appears likely that a stronger agreement could have been reached. It therefore appears that the United States could have overcome Malaysia's position but not *vice versa*. Of the two counterfactual propositions, only the second is plausible; it was not Malaysia's position on sovereignty that prevented effective agreement, it was the United States' own interests.

Aftermath

The outcome of UNCED represents the end of a process, but this did not bring to an end the push for a global forest convention. Since Rio there have been two general developments on the specific issue of a convention. First, despite the lack of language in the Forest Principles, and despite opposition from key countries, the question of a world forest convention has refused to die. Perhaps even more remarkably, the two key opponents in Rio have now seemingly reversed positions. Malaysia switched roles to become one of the leading proponents of a convention. Whether this shift in position represents a shift in interest in an *effective* agreement is questioned by some, given that a global convention is now pushed most strongly by industry and governments that have the greatest interest in timber exporting, such as Malaysia. Nevertheless, with the change in Malaysia's position around 1994 to active support for a convention, some analysts grew optimistic about the future of the debate. By 1996, Humphreys (1996b, 249) was able to argue that North-South unity was growing on the question of a forest convention. It was not recognized at that time that the United States was meanwhile shifting its own position toward firm opposition to a convention, nor that this single factor would be enough to prevent any further negotiation of a forest convention for more than a decade after Rio, right up to the present time.

The United States continued to support a convention, at least officially, until the mid-1990s.[130] Its turnabout was made official only in 1997, and came as a big surprise to those involved in global forest negotiations. However, US officials had already begun to rethink the US position on a world forest convention shortly after Rio:

> We were already thinking then, April 1993 maybe, me and others, that we're sort of in a convention fatigue. And we were beginning to think that negotiating these conventions was not about negotiating the protection or conservation of a resource, they were about new and additional resources, technology transfer, trade, the right to development, sovereignty, and almost not at *all* about the issue. . . . I think it was a process. It didn't happen overnight, none of this did; by 1993, 1994, certainly 1995. . . . But we sort of got really anti-negotiation. What do you get for it? You get a bunch of rhetoric, you know? . . . Forget those developing countries. If they didn't want it, wipe 'em off, you know? . . . We don't need to invest our time.[131]

Other explanations have also been offered for US opposition to a world forest convention in recent years. More than one source argues that the United States' position is very much based in acting on its own local laws and regulations—in other words, maintaining sovereignty—rather than "being on the receiving end" of international oversight such as a convention would bring. The limited role of the federal government over forests in the United States, due to a hybrid pattern of forest ownership, has also been noted.[132] One official asserts that this decentralized approach, which empowers people and industry, might be threatened by an international convention in that this might attract more power to the federal government in this area.[133]

One government official attributes US opposition to the belief that the United States "never wants to preclude future options. That's something that you find at all levels of US negotiators—you want the ability to change your mind. When you get into a legally binding agreement you eliminate that ability."[134] However, the United States has not always placed highest priority on this ability; as witness US leadership in pushing for the creation of the United Nations in the first place, or in the ozone agreements.

Another reason given is the lack of high-level political will for a forest convention in the US administration. In this the US position since UNCED differs radically from the early days when President Bush the elder went to bat personally with G-7 countries over the original US proposal. The official asserts, though, that this is a necessary but not a sufficient condition for success. It also begs the question of why the former president would go to bat for a convention while more recent ones would not.

Meanwhile, an NGO participant in Rio asserts that it is the influence of NGOs on the US position that has been decisive:

> After Rio, the US wasn't campaigning on a convention, but it clearly was open [to it and] had an interest in one, I think until NGOs coalesced and said they didn't want one. The idea of opposing a world forest convention was completely counterintuitive for NGOs. Global conventions are the highest manifestation of the rule of law; but it has to be the right kind of convention. . . . It took us years to discover that a convention will only have in it anything, or nothing, that the parties negotiating it want. It won't have what I want [as an NGO], I'm outside the process. I had to step back and think about the motivations of the countries wanting a convention. The ones who would negotiate it are Canada, Finland, Malaysia, and a few other [big timber exporters].[135] . . . It's very likely that the basket they wouldn't want in it would be my stuff! And all the stuff we didn't want was what they wanted, . . . such as ensconcing in international law a set of very weak standards that would thereafter become the test of

> whether forests were managed acceptably or not so that no one could ever protest again. And everybody would have to buy everyone else's timber and the forests would be deemed to be fine.[136]

It is noteworthy that NGO fears of too little regulation were the opposite of industry's stated fear of too much. In any case, there is currently little ambiguity regarding the US position on a global forest treaty: no constituency in the United States explicitly supports one. Therefore, whether consciously realized or not, opposing a convention is costless for the US government, and this is where the US official position remains.

Toward "A Legally Binding Instrument"?

US opposition to a forest convention across the board has had a huge impact on developments in global forest policy. This despite the fact that Malaysia and Canada and other states have engaged in strong and sometimes heavy-handed ongoing efforts to revive negotiations toward a binding agreement. The years since Rio have seen three successive international forest policy "talk shops"[137] established by the UN. The first, an open-ended[138] Intergovernmental Panel on Forests (IPF), was set up by the UN Commission on Sustainable Development (CSD), a body established in 1993 to monitor implementation of the UNCED Agenda 21 action plan. The IPF was mandated to review all Agenda 21 issues pertaining to forests before the fifth anniversary of UNCED in 1997 and to pursue consensus and develop coordinated proposals for action to support "the management, conservation and sustainable development of all types of forests," as stated in the Forest Principles. It was also instructed to discuss "the role of international institutions and legal instruments": an ambiguous term intended to reopen the question of a convention. After two years and four meetings, the IPF produced 140 *nonconsensual* "proposals for action," with no conclusions on the most contentious issues of financial assistance, trade-related matters, or a global forest convention.

At its "Rio + 5" UN General Assembly Special Session (UNGASS) in 1997, it was agreed that the intergovernmental forest dialogue would be continued through a new *ad hoc* body, the open-ended Intergovernmental Forum on Forests (IFF), which would report to the CSD in 2000. Language on a possible convention was beefed up slightly in the UNGASS Decision that the Forum would also "identify possible elements of and work towards a consensus for international arrangements and mechanisms, for example, a legally binding instrument."[139]

The IFF carried a sense of "IPF take two" with it, in part because the role of the IFF—like that of the IPF—was unclear,[140] and in part because the struggle over a forest convention lurked in the background, underlying consideration of all other issues (Davenport et al. 1998, 11). Language reflecting G-77 demands for increased financial flows and technology transfers—and developed country resistance to making any concessions—also changed little between the IPF and IFF (Humphreys 1996b, 161). This time, however, the coalitions began to solidify along different lines. Canada took a forceful lead on the issue as the United States backed away from it. The pro coalition also included Switzerland, Russia, Poland, Chile, Malaysia, Senegal, Panama, Guatemala, and Costa Rica. The anti coalition comprised the United States, on record as opposing a convention after 1997, Cuba, New Zealand, Mexico, Ghana, Japan, India, and Brazil, along with the other Amazonian countries. The EU was formally bound to support a forest convention, and did so, but did not actually have a unified position after the proposal to begin negotiations on a convention failed to win consensus in the IPF.[141]

The IFF continued the dialogue in much the same fashion of the IPF, adding a further 130 nonconsensual "proposals for action" to the 140 produced by the IPF. At the final session of the IFF the question of follow-up was tackled. The United States proposed to institutionalize the forum for discussion and the United Nations Forum on Forests (UNFF) was born. It was placed under the aegis of the UN Economic and Social Council (ECOSOC) itself, one of the six primary organs of the UN, in order to give it the political authority to address issues associated with the IMF, the WTO, the World Bank and other high-level institutions. The UNFF was also given a sister body, the Collaborative Partnership on Forests (CPF), composed of representatives of organizations or institutions that have an impact on forests, such as the FAO, the ITTO, and the Convention on Biological Diversity (CBD).

The negotiations for the new UNFF and CPF were not straightforward, however. Indeed, the process was pushed to the brink of collapse when Canada, which had constantly been the strongest advocate of a forest convention since the end of UNCED, asserted that it could not support the creation of the UNFF if there was no consensus for a convention. This was "a high-risk tactic," in Humphreys' words, as Canada not only bracketed all references to the UNFF but also refused to negotiate on finance until other states agreed to a convention. Ultimately, delegates agreed that ECOSOC and the UN General Assembly would "within five years, and on the basis of [an] agreed-to assessment of the arrangement, consider with a view to recommending the parameters of a mandate for developing a legal framework on all types of forests."[142] At the same time, this process could develop the financial provisions to implement any future agreed legal framework, and the General Assembly and ECOSOC would "take steps to devise approaches towards appropriate financial and technology transfer support."[143] Given that these compromises satisfied neither those who wanted a convention nor those who demanded increased forest-related aid, the United States was one of very few countries that actually fully supported both (Humphreys 2001, 164). The IFF and the CSD passed the actual decision on where the UNFF should be situated within the UN system on to the ECOSOC, which institutionalized the UNFF as one of its own subsidiary bodies.

The stated purpose of the UNFF, unlike that of the IPF and IFF, was primarily to promote implementation, in particular, the implementation of the IPF/IFF proposals for action. Yet the UNFF was not given a mandate to "specify what policy targets countries should pursue, when to achieve them, how to finance them, or to whom and when to report the results" (Dimitrov et al. 2001, 12). Even if it could do that, it would have no way to enforce compliance in meeting any targets.

Some observers suspect that some of the limitations of the UNFF may at least partially reflect a deliberate attempt by some pro-convention delegates to undermine the process, thereby giving impetus to the only remaining alternative, a legally binding convention on forests (Dimitrov et al. 2001, 12). Yet its structure and purpose were based on a US proposal[144] and reflected almost entirely the United States' wishes. In answer to whether the United States was happy with the end result, one official noted in 2001, "I think we got a lot of what we wanted with how the UNFF was created; we won a great deal actually. . . . We're pretty happy with it, frankly."[145] This is not surprising, given another US official's comment that "the US generally gets its way",[146] although, again, what the United States wants depends on the costs and benefits, as demonstrated by the Rio outcome on forests.

A short time later, however, some US officials were speaking of the "inevitability" of a convention.[147] Some observers pointed out that a convention with low standards could provide a way for industry to avoid the high standards that must be met in some of the more popular private programs for certifying timber and timber products as "sustainably produced" while still claiming to be environmentally conscious.[148] The adoption of a low global standard in a binding agreement was also thought to be attractive to industry in that it could influence the United States to abandon stricter domestic regulations, as happened more than once in the ozone case:

> I think they [the US industry] have finally come around to the Canadian-Malaysian position, that standards have nowhere to go but *down*, if you go global. The thinking before was that it would bring everybody else up, and that that would help us. Now they're saying it'll allow us to go down.[149]

At that time, while not confirming such rumors, an industry spokesperson told me that it would not be difficult to change the United States' position to match its own:

> We think we can bring over the US administration to favor a world forest convention [if we decide we want one]. You carve off the moderate NGOs, like the Conservation Society; once you get their support for a convention you can influence the government.[150]

By the time of the mandated review of the UNFF at its fifth meeting in 2005 (UNFF-5), even the United States, the body's staunchest supporter, acknowledged a need for improvement (Baldwin et al. 2005, 4). It was part of a small group of interested countries that came to the meeting with a proposal for an improved "international arrangement on forests" (IAF), to follow on from UNFF, which for the first time would incorporate a funding mechanism. This proposal for funding differed from the "Forests for the Future" fund that the United States proposed in 1992 in that this would be a "seed fund" established by intergovernmental members of the CPF, such as the FAO and the World Bank. It was interesting in that it showed that the United States' main concern was not the question of additionality, at least not limited additionality. Indeed, it indicated that the United States was prepared to offer an incentive to try to ensure that countries continue to support nonbinding arrangements. It has been suggested that this may be advantageous to the United States too as it provides a veneer of multilateralism while allowing it the flexibility to pursue perhaps more unilateral interests.[151]

By the end of UNFF-5, delegates had come to no agreement on the future path of the IAF. The discussions were abruptly halted and were not concluded until UNFF-6, in February 2006, with an agreement to keep talking (Hajjar et al. 2006).

What of American industry? During the process of preparing for the UNFF review in 2005, rumors again abounded that the industry association in the United States had finally decided to support a global forest convention. This was denied by an industry spokesperson, however,[152] and the United States maintained its opposition to a convention. In any case, even if the United States forest and forest products industry were to push for a global forest convention, the question remains whether any convention advanced as a result would be truly effective according to the definition offered here.

There is a palpable sense of frustration on all sides with the lack of concrete environmental results emanating from the dialogue of over almost a decade and a half, and this frustration is likely to persist as deforestation continues apace. On the other hand, it has not led to an increase in support for a convention. One overwhelming sense at recent UNFF meetings was that the push for a forest convention has lost steam.[153] Certainly the United States still opposes it, and, indeed, unless that position were to change, I would predict little success for those countries still advocating it. Many of its advocates seem resigned that it will not happen any time soon.

US Interests: A Breakdown

The United States' proposal and efforts to lead the world toward a global forest convention in the early 1990s, together with its subsequent rejection of further efforts toward a convention, form an intriguing riddle. Can one explanation enable us to understand both stances? Yes, if we examine the net stream of benefits and costs that the United States could expect to incur from agreement. Such an analysis shows that the same conditions that made it impossible to achieve an effective binding agreement in 1992 also explain why there has been no change in that outcome ever since.

As outlined in chapter 2, possible benefits may fit into any one of three categories: the avoidance of environmental damage, the avoidance of costs due to loss of market competitiveness, and any positive, particularistic benefits that may accrue to some actors as a result of environmental regulation. Costs may include the cost of research, development, and marketing of any potential substitutes for regulated substances or products, the cost of halting economic activities that depend on the regulated or banned activity or product, and any transaction costs entailed in offering incentives (positive or negative) to coerce other states into an agreement. When perceived costs and benefits to the United States are broken down, a clear relationship can be seen between US interests and the lack of a global forest convention today.

Environmental benefits: With regard to environmental benefits, the United States initially had some interest in the benefit of avoiding the environmental costs of deforestation, given the fact that tropical deforestation was occurring at an unprecedented rate. The US proposal would encourage, if not require, action on the part of other countries, mainly tropical forest countries, and US policy-makers expected this to produce an improvement over the status quo. While some of the environmental benefits would be felt only at the local level, particularly in countries with tropical forest, others were more global and would benefit the United States as well. Two global forest benefits of particular salience to the United States were the environmental services of carbon storage and biological diversity.

After Rio, other processes and initiatives were set in motion, including numerous regional processes to define criteria and indicators for sustainable forest management and other processes that aimed to certify timber as coming from sustainable sources. Although deforestation had not been halted by any means, these international endeavors had the effect of reducing somewhat the sense of urgency over the potential threat of environmental damage from deforestation, as the perception began to take hold that the problem was being addressed. Thus the perceived benefit of avoiding environmental damage through a global forest convention was also slightly diminished.

Avoidance of economic costs: The US government initially placed some value on the prospective benefit of avoiding competition from timber sources that did not have to meet environmental standards as high as those in the United States. However, the industry itself was somewhat divided on this, for several reasons that have been detailed above. Leveling the playing field would be of interest to US timber producers and from an economic angle this was probably the most significant benefit sought. However, the more powerful timber processing branch of the industry did not stand to benefit as much, and potential benefits had to be balanced against the threat that industry might be faced with increased costs from higher international standards or international oversight. This perceived threat seemed to increase with the change in US administration in 1992. Meanwhile, the potential benefits of raising standards for all timber producers were diminished somewhat as niche markets for certified and labeled wood products meeting higher environmental standards began to open up in the mid-1990s.

Positive economic benefits: The possibility of a particularistic positive benefit for US industry was not in evidence in the forest domain during preparations for UNCED and this still holds true. The proposed forest convention offered US producers neither a competitive advantage over their competitors nor the prospect of greater profits from new methods of timber production or processing. Nor would timber producers benefit from innovations in timber substitutes that a convention might promote.[154]

For these reasons, and whatever effects that inexperience with international forest issues may have had, the US forest industry was not a major proponent of a forest convention, even during the period when the United States was attempting to push forward a global convention. This lack of industry interest indicates that the total expected benefit of a convention was fairly low for the United States during the UNCED process and even lower afterward.

One other benefit existed that was unique to the circumstances of this case: the benefit to be gained—at the proposal stage—from "killing" the protocol on energy use that was being proposed for the climate change convention then under negotiation. Once that goal was accomplished, there were no overwhelming benefits to be gained for the United States.

Costs of substitutes/halting activities: As to costs, there were few costs to the United States in its initial proposal because the proposal did not call for action on its part, apart from a vague reference to providing technical assistance. This is to be expected, as countries usually propose what they themselves are already doing, as the least costly or disruptive to themselves,[155] just as states are more likely to adopt regulations that have a low cost (DeSombre 2000, 232). The two major costs of taking action in the United States—the cost of substitutes and the cost of halting or restricting economic activity—arose neither during the Rio process nor at any time since then. Substitutes for timber in the United States were not proposed and were never debated, nor was there talk of halting any economic activity (specifically, timber production) in the United States that depended on harvesting trees, despite the Pacific Northwest controversy over old-growth logging that was taking place at that time.

Certainly the Forest Principles as finally agreed held no real cost implications for US industry. Heissenbuttel et al. produced a joint paragraph-by-paragraph commentary on the Forest Principles shortly after UNCED. Assuming full implementation of the Forest Principles and using costs to the industry as primary indicator, Heissenbuttel, of the AF&PA, predicted that of 43 principles (including sub-paragraphs), 17 would have no significant impact on US forestry and 16 would have some kind of positive or favorable impact; his most negative prediction was that the impact of six of the principles would "depend on interpretation and application" (Heissenbuttel et al. 1992).

Cost of manipulation: In the end, it was the third type of potential cost, the cost of manipulating other states into agreement, that tipped the scales in favor of no binding agreement (nor any mention of the future possibility of a binding agreement) at Rio. If, as the United States argued, the issue of sovereignty had been paramount, then it is true that no amount of money, nor any other *quid pro quo*, could have bought the South's cooperation. The numerous indications of willingness to bargain that were given by Southern states during and between negotiation sessions contradict this view, however.

The United States' interest in using its resources to manipulate the preferences of the anti coalition—as determined by the shifting net value to the United States of a convention—was not great enough to allow for effective leadership. This was exacerbated by the relatively high value of no agreement for the anti coalition in terms of sovereignty over their forest resources and the very low or nonexistent value of agreement to the anti coalition if it were not linked to meeting other value-claiming demands. Because the United States lacked a clear interest in an effective forest convention, based on assessment of the costs and benefits to itself, it made few efforts to shift the costs and benefits of effective agreement for other actors. The paucity of leadership on the part of the United States thus not only links back to the United States' economic assessment of interest but also forward to the lack of an effective outcome in this case.

The real factor behind the lack of an effective binding agreement on forests, or even a commitment to future negotiation of a binding forest agreement, thus lies in the cost of leadership itself for the United States. Arguably, the United States was ready to bear some small cost for a forest convention, as witness its (relatively costless) support for a desertification convention and the $150 million for forest conservation in developing countries that President Bush eventually offered in Rio. While these may well have helped the United States to win a developing country commitment to sign the Forest Principles and secured some extremely vague preambular language on possibilities for future cooperation, they were not enough to change the anti coalition's valuation of agreement relative to the no agreement outcome; costlier actions were needed. Leadership by the United States—and thus the achievement of any binding convention, whether it would meet my criteria of effectiveness or not—will ultimately depend on whether this type of cost can ever be balanced out by benefits for the main US domestic actors involved.

Appendix 5.1: Timeline of Global Forest Policy Developments

1990 Proposal for a world forest convention made by United States at G-7 meeting: July (Houston)
PrepCom 1 for UN Conference on Environment and Development (UNCED) August (Nairobi)

1991 PrepCom 2: February–March (Geneva)
PrepCom 3: August–September (Geneva)

1992 PrepCom 4: March– April (New York)
UNCED: June (Rio de Janeiro)

1993 First meeting, Commission on Sustainable Development (CSD): May–June (New York)

1995 CSD establishes the *ad hoc* Intergovernmental Panel on Forests (IPF): April (New York)

1997 IPF reports to CSD: April (New York)
UNGASS establishes the *ad hoc* Intergovernmental Forum on Forests (IFF): June (New York)

2000 IFF reports to CSD: April (New York)
ECOSOC establishes UN Forum on Forests (UNFF) and the Collaborative Partnership on Forests (CPF): September (New York)

2005/6 UNFF review of the international arrangement on forests: May 2005, February 2006 (New York)

CHAPTER 6

The Climate on Climate Change

As I write, the city of New Orleans is under water for the second time in less than one month, and perhaps as many as 1,000 of its inhabitants are dead from Hurricane Katrina and its aftermath. Hurricane Katrina is being called the costliest natural disaster ever to hit the United States.[1] But it has also been labeled the worst "unnatural" disaster,[2] as human decisions are thought to have played a major role in increasing the human toll, especially in New Orleans.

It is not just the appropriateness of government responses to the disaster that is in question; the debate also encompasses the human causes behind the occurrence of the disaster itself. The sapping of ground water has been blamed for subsidence of the city ever further below sea level,[3] while engineering of the Mississippi River basin in order to prevent floods and aid navigation has led to the loss of more than 1,500 square miles of Louisiana's coastland in the last 50 years.[4] One of the issues raised in the media and in scientific circles, however, concerns human activities with less direct effects but perhaps just as lethal. A long-term human-induced change in the global climate is now thought by some researchers to be implicated in increasing the severity and frequency of extreme weather events such as the Category 5 Hurricane Katrina.[5]

The United States, however, has never supported concrete global commitments to reverse its continually growing use of fossil fuels and the resultant greenhouse gas emissions that are at the root of the climate change phenomenon. An analysis of the history of climate change negotiations and a breakdown of the domestic interests that are potentially affected explains why opposition has, up to now, prevailed within the United States.

The Phenomenon of Global Climate Change

The theory of global warming is much older than the theory of ozone depletion, having first been postulated in 1896 by a Swedish chemist, Svante Arrhenius. The "greenhouse" metaphor had already been used as early as the 1830s to describe the effect of the earth's atmosphere on its surface, but it was Arrhenius who first calculated that human activities such as the burning of fossil fuels might alter the natural climate (Schröder 2001, 10).

The sun radiates infrared light on the earth. The earth's atmosphere is associated with a greenhouse because the gases within it act like a greenhouse in delaying the escape of infrared radiation back into space. Naturally occurring greenhouse gases (GHGs), including water vapor and carbon dioxide, make up about 1 percent of the earth's atmosphere and are in fact what makes life possible on earth, trapping the heat of the sun and keeping the earth as much as 30 °C warmer than it would otherwise be. However, anthropogenic—or human-made—GHGs, such as carbon dioxide emitted from fossil fuel burning, methane and nitrous oxide produced by farming activities and changes in land use, and several industrial gases that do not occur naturally, have thickened the natural blanket of GHGs in the atmosphere with unprecedented speed. This is the "enhanced" or "anthropogenic" greenhouse effect.

There is an almost universal scientific consensus that anthropogenic GHG emissions are likely to warm the surface and lower atmosphere of the earth. Such global warming is, in turn, predicted to cause changes in cloud cover, precipitation and wind patterns, and the duration of seasons—in other words, climate change.[6] Indeed, there are numerous signs, such as possible increases in the intensity of hurricanes as mentioned above, that some of these effects may already be occurring. Concentrations of GHGs in the atmosphere were estimated in 1992 to be some 25 percent over their pre-industrial levels (Stone 1992, 446); it is currently predicted that they could double or even triple their pre-industrial levels during the twenty-first century.[7]

The Global Climate Regime and the Criteria of Effectiveness

The UN Framework Convention on Climate Change (UNFCCC), signed at the UN Conference on Environment and Development (UNCED) in Rio de Janeiro in 1992, and its 1997 Kyoto Protocol form the two binding agreements that I here label the "global climate regime."[8] Intended to address the threats that climate change may pose to humans, this regime has shown itself to be almost entirely ineffective in mitigating the problem despite the entry into force of both the Convention and, in 2005, the Protocol that lays out specific commitments to behavioral change. Once again, using the table presented in chapter 2, we can easily discern the specific areas in which the climate regime lacks effectiveness.

Commitments: The negotiators of the Kyoto Protocol followed the Montreal Protocol model in their attempt to agree on specific reductions to be made in emissions of six GHGs—carbon dioxide (CO_2), methane (CH_4), nitrous oxide (N_2O), hydrofluorocarbons (HFCs), perfluorocarbons (PFCs), and sulfur hexafluoride (SF_6)[9]—with a specific timetable for achievement over the period 2008–2012. The general commitment expressed in the Protocol is for the agreeing parties to reduce their overall emissions of GHGs by at least 5 percent below 1990 levels for the commitment period 2008 to 2012, or "first commitment period" (Article 3.1).[10]

This commitment is weak in numerous ways. First, it is very small when compared with scientific assessments of what is required. Even before UNFCCC negotiations began, scientists and representatives of NGOs, intergovernmental organizations, and governments agreed in Toronto in 1988 that a 20 percent reduction of CO_2 emissions—which alone comprise 60 percent of all GHG emissions[11]—from 1988 levels was needed by the year 2005, with a full 50 percent cut needed in the long term (Schröder 2001, 14).[12] Second, this commitment was made only by "Annex I countries," that is, countries listed in the first annex of the UNFCCC itself, which includes developed countries and Eastern European countries "in transition" to market economies. The list did not include developing countries, meaning that this, the most concrete commitment of the regime, does not apply to them.

Moreover, each country was allowed in Kyoto to make commitments to achieving *individual* "quantified emissions limitations and reductions objectives" (QELROs) (Bettelli et al. 1997, 1). These would be achieved through ensuring, individually or jointly, that their aggregate anthropogenic carbon dioxide equivalent emissions of GHGs do not exceed individually assigned amounts, listed in an annex of the Protocol, during the commitment period 2008–2012 (Article 3.1). In other words, each country chose for itself the amount of GHG emissions reductions that it would attempt to achieve on average over the period 2008–2012. Many chose targets in the 7 to 8 percent range, but Australia, Iceland, and even Norway were willing to commit themselves only to slowing growth in emissions, rather than to actual reductions.[13]

In addition, the Protocol allows for flexibility in achieving these targets. For instance, 1995 may be used as the base year for HFCs, PFCs, and SF_6 (Article 3.8). Calculation of changes in emissions is also to take account of changes in GHG removals by sinks resulting from direct human-induced land-use change and forestry activities—afforestation, reforestation, or deforestation—since 1990 (Article 3.3). "Supplemental" emissions trading between parties is also allowed under Article 17. Transfer of emission reduction units resulting from projects aimed at reducing emissions or enhancing removals of GHGs by sinks is also allowed between Annex I parties.

Compliance: Article 5 of the Protocol lays out instructions for estimating anthropogenic emissions and global warming potentials of the six listed GHGs by sources and removals by sinks according to methodologies accepted by the Intergovernmental Panel on Climate Change (IPCC) and agreed upon by the Conference of the Parties (COP) to the UNFCCC (UNFCCC Article 5). Parties are then to report on implementation in national communications and provide supplementary information necessary to demonstrate compliance (Article 7). The information provided is to be reviewed periodically by a review team selected from experts nominated by the parties and intergovernmental organizations (Article 8). This team is to assess implementation and identify any potential problems in the fulfillment of commitments and circulate its reports to the parties to the Convention via the secretariat. Questions of verification of national reports and procedures for enforcement of compliance were left for further elaboration at a later date (Articles 17 and 18).

In November 2005, the first Meeting of the parties to the Kyoto Protocol (COP/MOP-1) adopted a decision setting out compliance procedures and calling for consideration of an amendment on compliance. The decision establishes a compliance committee to promote, verify, and enforce compliance through such possible mechanisms as suspending a member's eligibility to participate in flexibility mechanisms or reducing its assigned amount of emissions during the second commitment period (post-2012) by 1.3 times its amount of excess emissions during the first commitment period.[14] As Black points out, however, this sanction is meaningless if there is no second commitment period; a second commitment period is still to be negotiated.[15]

The UNFCCC itself establishes a procedure for settlement of disputes (UNFCCC Article 14), which also applies to the Protocol (Article 19). It allows parties to accept compulsory submission of a dispute with any party accepting the same obligation to the International Court of Justice and/or arbitration under procedures to be adopted by the COP (UNFCCC Article 14.1–2). The Convention otherwise recommends peaceful means of settlement and allows for one party to a dispute to request creation of a conciliation commission to recommend an award for the parties to "consider in good faith" (UNFCCC Article 14.6). Other procedures for conciliation were left for later COP decisions (UNFCCC Article 14.7).

Mechanisms to assist with implementation are laid out in both the UNFCCC and the Kyoto Protocol. The UNFCCC refers to the establishment of a "Financial Mechanism" the operation of which "shall be entrusted to one or more existing international entities" (UNFCCC Article 11.1). This represented a victory for donor countries, listed in Annex II of the Convention, who wanted to refer funding questions to the already-established and World Bank-controlled Global Environment Facility (GEF) (Bodansky 2001, 33), although the GEF was not accepted as the permanent funding mechanism until 1998 (Carpenter et al. 1998). The Protocol, for its part, establishes the Clean Development Mechanism (CDM), to assist non-Annex I parties in achieving sustainable development and in contributing to the ultimate objective of the Convention, and to assist Annex I parties in achieving compliance with their QELROs commitments (Article 12.2). Unlike the Multilateral Fund for Implementation of the Montreal Protocol (MLF) under the Montreal Protocol, it is not a fund, although it can assist countries needing assistance in arranging funding for project activities "as necessary" (Article 12.6). However, the Protocol envisages that funds will come from other parties, which would then be allowed to use the (certified) emission reductions accruing from such project activities toward compliance with part of their own QELROs commitments under Article 3 (Article 12.3 (b)).

Participation: Participation in the Framework Convention is almost universal, with 189 ratifications.[16] However, the UNFCCC contains no concrete commitments, apart from a commitment to report on national inventories of anthropogenic emissions by sources and removals by sinks of all greenhouse gases (UNFCCC Article 4.1(a)). The Kyoto Protocol itself has 163 parties as of April 18, 2006, having entered into force in February 2005. The large number of parties, however, belies the fact that Annex I countries that are parties to the Protocol represent only 61.6 percent of total emissions from Annex I parties.[17] As is well known, the United States is not a party, although it is the largest emitter of carbon dioxide, producing about 22 percent of global emissions in 1994. But the second and fifth largest CO_2 emitters, China and India, which produced 13.4 percent and 3.8 percent of global CO_2 emissions in 1994, respectively, are not covered by the Protocol's commitments to reductions either, as they are not Annex I parties.[18]

Treaty relationship: Because some gases covered by the Montreal Protocol are also greenhouse gases, the Kyoto Protocol specifies that it covers only emissions and reservoirs of greenhouse gases not controlled by the Montreal Protocol (Article 2.1 (ii) and (vi)). It also takes into account each party's commitments under relevant international environmental agreements (Article 2.1 (ii)). It thus specifies a subordinate position in relation to the Montreal Protocol and other treaties under which any party has made commitments.

Table 6.1 shows how the climate regime scores on effectiveness according to the individual subcriteria of effectiveness identified in chapter 2.

Table 6.1 The Global Climate Regime and the Criteria of Effectiveness

Commitments	*Compliance*	*Participation*	*Treaty relationship*
precise targets 2	specified monitoring or review procedures 2	producers of the problem 1	language specifying relationship to other instruments 1
timetables for their achievement 2	body established to review implementation 2	potential providers of financial resources to assist implementation 1	superiority to other instruments 0
firm, concrete commitments rather than "indicative goals" 2	specified dispute resolution body 0	potential providers of technology to assist implementation 1	specified procedure for resolving conflicts with other instruments 0
	enforcement procedures 1*	universal participation 0	specified body to resolve conflicts with other instruments 0
	sanctions for non-compliance 1*		
	mechanism for assisting implementation 1		

Coding:
0 = nonexistent
1 = somewhat effective
2 = effective

* This is scored as "1" given the adoption of a compliance mechanism by decision at COP/MOP-1 and the ongoing consideration of an amendment on a (very weak) compliance mechanism which COP/MOP-1 set in motion in 2005.

The regime scores highly on commitments, because commitments are made to achieving precise targets for reductions in GHGs in the atmosphere and there are stated timetables for their achievement. These scores, however, mask the fact that the targets are weak in terms of the actual reductions in emissions that are required. On compliance, reporting and review are specified, but there are no specified procedures for settlement of disputes or enforcement of compliance, nor are sanctions for noncompliance addressed except through a COP/MOP-1 decision and ongoing efforts to amend the protocol. The CDM does exist to assist compliance, but its function is to coordinate between parties seeking assistance and Annex I parties willing to contribute funding for projects in return for credit toward meeting their own commitments. The Protocol scores even lower on participation, encompassing concrete commitments toward actual reductions of GHGs by only 36 of the 189 countries that ratified the original UNFCCC, representing only 62 percent of all Annex I parties' emissions. Three countries that are by themselves responsible for roughly 40 percent of all GHG emissions globally—the United States, China, and India—are under no commitments to reduce their emissions, and because the United States is not a party its resources are not available for financial or technological assistance. Finally, the Protocol is subordinate to those treaties to which it specifies its relationship.

The Negotiation of the Global Climate Regime

In 1987, scientists gathered at meetings of experts in Villach, Austria and Bellagio, Italy, called for an examination into the possible need for a convention to address global warming. It was not just a growing scientific consensus that the threat of climate change existed that prompted action. According to Schröder, it was no coincidence that the call was made during an era of heightened environmental awareness in many industrialized countries that forced "green politics" onto the political agenda. This environmental movement was prompted by several factors, including Chernobyl, the Exxon Valdez spill off Alaska, and several extreme weather events in the 1980s, including the fact that the 1980s was up to that time the hottest decade on record. Although not all of these environmental events were related to climate change, they "fuelled . . . a growing sense of insecurity about the climate" (Schröder 2001, 17). As mentioned in chapter 1, the political response to climate change was also influenced by the successful outcome of the ozone negotiations, and the Vienna Convention and its Montreal Protocol were quickly taken up as a model for climate change negotiations.

Intergovernmental efforts to mitigate climate change started up shortly after the meetings in Villach and Bellagio, with a call by ministers for the adoption of targets or national strategies to limit GHG emissions at the Second World Climate Conference which took place in late 1990. Negotiations were formally launched with the creation of the Intergovernmental Negotiating Committee on Climate Change (INC) by the UN General Assembly in December 1990. More than 150 states came together to form the INC, which met five times before UNCED in 1992.

Although the INC was mandated to draft a framework convention on climate change through broad consensus, it dealt with contentious issues including targets and timetables for reducing CO_2 emissions, the "common but differentiated" responsibilities of developed and developing countries, and the need for financial and technological transfers to developing countries (Carpenter, Chasek, and Wise 1995, 1). Paterson and Grubb identified several major fault lines within the group of negotiating states that came to light as the early negotiations progressed (1992, 295–297):

- a North-South divide, particularly over how to share the burden of GHG reductions and how to assist developing countries in meeting their burden through financial and technological transfers;
- a split between producers and exporters of oil and coal and those who rely on imports of fossil fuels, over the question of reducing use of fossil fuels in order to reduce GHG emissions;
- a split between states that are relatively more resilient to the impacts of climate change and those for whom the issue is more salient, in terms of vulnerability to either extreme weather or rising sea levels and/or in terms of a lack of economic resources needed to adapt to the impacts of climate change.

Given these rifts, there is no doubt that while the goal of a convention loomed large in the negotiations, some countries desired a less than effective convention, and that others preferred none at all unless the value of a convention to them were raised. Once again, then, a deadlock existed.

The United States in Early Negotiations

The United States has long been the world's largest emitter of GHGs, responsible for about 25 percent of total global emissions, in volume of CO_2 equivalents (Victor 2004, 42). Even before the INC was formally established, the United States' position was noticeably at odds with that of the ten European governments that pushed to begin negotiations. At a UNEP-sponsored ministerial meeting in the Netherlands in 1989, the United States, along with three other states, prevented agreement on an informal commitment to stabilize CO_2 emissions by the year 2000. The United States instead focused on the need to improve knowledge regarding the science behind global climate change (Monastersky 1990a, 102).

The United States tried to downplay differences. William Reilly, the US Environmental Protection Agency administrator, insisted that the United States supported a reduction of carbon dioxide although he was not prepared to say by when and by how much. Later he downplayed US negativity in comparison to the lack of any indication by the USSR and China—the second and third largest emitters of CO_2 at that time—of readiness to cut emissions (Monastersky 1990b, 263). A leaked "talking points" memo to members of the US delegation at another small conference in April 1990 also warned that it was "not beneficial to discuss whether there is or is not warming, or how much or how little warming".[19]

By the beginning of the first INC less than a year later, the United States had become a more open opponent of significant international reductions in GHG emissions, even while many European countries were adopting unilateral targets for GHG emission reductions (Haas 1992b, 2). However, there was still some optimism. Haas predicted that "neither the US' nor other countries' constituencies will tolerate any longer the US administration's invocation of economic and scientific justifications for inaction" in its efforts "to retard progress on a global climate change treaty" (1992b, 3). Certainly Mostafa Tolba, UNEP's executive director who had been given much credit for pushing the Montreal Protocol negotiations toward their successful conclusion, stood ready to play a lead role once again and push states to make commitments on targets and timetables, this time for reducing GHG emissions.[20] In the case of climate change, however, because these aims did not coincide with those of the dominant state, Tolba's call went unheeded until 1997 and even beyond as the United States distanced itself from his views.

Meanwhile, Haas's general prediction proved inaccurate, as other negotiators found there was little they could do except indeed "tolerate" the United States' efforts to retard progress. The United States steadfastly refused to consider language setting quantitative targets to stabilize CO_2 emissions—much less reduce them—and was "markedly more hostile than most . . . other OECD countries to proposals for significant North-South transfers" (Paterson and Grubb 1992, 302). As a result, the negotiations moved slowly. By the end of the first two-week session of the INC, negotiators had only "negotiated on how to negotiate."[21] This snail's pace continued to the end, due to obstruction by various countries that together had the power to hold progress back on achieving an effective consensus, including the United States.

Negotiations on the Framework Convention on Climate Change ended in May of 1992 with no agreement on targets for GHG emissions reductions or timetables for achieving them. The final blow came just ten days before the final adoption of the Convention, when then-British Environment Minister Michael Howard agreed to back the United States' demand to remove bracketed text (text proposed but not agreed) in the draft convention setting a goal for stabilization of GHG emissions at 1990 levels by the year 2000 (Weisskopf 1992, A1). The final compromise text instead refers to returning to "earlier levels" by the end of the decade and calls for a review of the adequacy of commitments at the UNFCCC's first Conference of the Parties (COP-1) (UNFCCC Article 4.2 (d)). The failure to get a commitment on targets and timetables exemplifies just how difficult it is for other countries to pull the

United States into an accord that runs counter to its perceived interests, and demonstrates how strong the United States' veto is even when United States is at odds with the rest of the industrialized world, as it bucked an almost unanimous consensus on the part of other developed countries favoring specific commitments (Weisskopf 1992, A1). This time the US administration did not attempt to use scientific uncertainty as a rationale; instead, US delegates cited economic uncertainty about the cost of reaching such targets.

Along with this ineffectiveness on specific commitments, the Convention was limited in other ways, particularly by weak compromise language that was used to paper over a North-South divide that had been exposed during the negotiations. The South, once again, had little interest in making commitments to address what it saw as a Northern problem. The North had created the problem in the first place with its fossil fuel-based industrial production since the Industrial Revolution. Per capita emissions in the South were far below those of industrialized countries—only one-tenth of the OECD average (Paterson and Grubb 1992, 297). More generally, many Southern countries saw poverty alleviation as their main priority and economic development as the mechanism to achieve this—the very economic development that would be threatened by any commitment to reducing emissions produced in the course of developing. Most developing countries, apart from those that were more vulnerable to increases in extreme weather events or rising sea levels, such as the small island states, were only interested in an effective agreement if it would grease the wheels for substantial transfers of funding and technology to assist them in developing without increasing emissions (Paterson and Grubb 1992, 297).

This was related to issues of equity: It was seen as grossly inequitable that developing countries should commit to lowering emissions, particularly in countries where this might detrimentally affect a state's sovereign right to exploit resources found within its own territory, such as the "dirty" coal of China (Higgins 1992, 8), after developed countries had used the same dirty technologies for their own development and had thus caused the problem. Developing countries were also not unaware of the fact that there were no concrete commitments for developed countries included in the Convention because the United States had virtually single-handedly removed those that other developed countries had sought to include, for economic reasons. Given all these developments during the negotiations, why indeed should developing countries make any commitments under the Convention?

Potential donor countries were not willing to acquiesce in the South's demands for funding and technology transfer. It would not be fair to characterize developed countries on the whole as unsympathetic to the arguments of developing countries, however. They included in the UNFCCC an acknowledgment of "common but differentiated responsibilities and respective capabilities" and the principle that developed countries "should take the lead in combating climate change" (Article 3.1). It was also true that many donor countries were willing to contribute financial resources to aid developing countries in working toward the goals of the Convention, but in many people's eyes this need had already been addressed sufficiently, assuming that the three-year GEF pilot project initiated in 1990 were to be made permanent. The developed countries therefore did the only thing that could be done to bring the developing countries on board the Convention, given the United States' unwillingness to bear the economic burden of either commitments on reductions or promises of additional financial or technological transfers: they accepted extremely weak commitments for developing countries.

In this they followed the model of the Montreal Protocol, establishing different categories of parties to the Convention. Those in Annex I were industrialized countries that committed themselves to "take measures on the mitigation of climate change, by limiting . . . emissions of GHGs and protecting . . . GHG sinks and reservoirs" in order to "demonstrate that developed countries are taking the lead in modifying longer-term trends in anthropogenic emissions consistent with the objective of the Convention" (Article 4.2(a)). Those countries listed in Annex II of the UNFCCC were the donor countries (and the EC) that committed to "provid[ing] new and additional financial resources to meet the agreed full costs incurred by developing country Parties in complying with their obligations" to provide a national inventory of anthropogenic GHG emissions and removals and other information (Article 12.1). They were also the donors to the international financial entity that would provide financial resources "to meet the agreed full incremental costs" of implementing any developing country's measures to mitigate climate change, once agreed between that party and the international entity.

It was envisaged by developed countries that the role of this "international entity" would be taken by the already-established GEF, as mentioned above. This did not please developing countries. Given that the GEF was already in place, this language added little. Moreover, the same issues the developing countries had raised with regard to the developed countries' domination of the GEF were still at issue in June 1992. Furthermore, the funding offered by GEF was not enough to do what it was intended to do: to meet the "full" agreed incremental costs of developing countries' actions to mitigate climate change. As Robert Watson, chair of the IPCC, pointed out in 1997, tackling global warming would require "the transfer of tens of billions of dollars" into the economies of developing countries in order for them to develop using new GHG technologies,[22] yet as of February 1992 the GEF had been allocated no more than $1.3 billion per year for grants and technical assistance.[23]

The result of the measures taken to resolve the North-South disputes, then, was that the Convention as finally agreed did not win the hearts and minds of developing countries. It also offered little to the United States except annoyance, given that any benefits the United States might gain from attempting to mitigate climate change were unquestionably outweighed by their costs. Finally, it offered little to those who had an interest in an *effective* convention. The negotiators, as Barrett would describe it, exchanged depth for breadth.

The UNFCCC was never intended to be the end of the story, though. The UNFCCC was opened for signature at UNCED, where it received 155 signatures. Practically before the ink was dry on the Convention, moves were afoot to begin negotiations on a Protocol—again following the Montreal Protocol model—that would incorporate real commitments. These moves included keeping the INC alive until the entry into force of the UNFCCC, in order to keep up momentum (Bodansky 2001, 34).

The Negotiation of the Kyoto Protocol

The UNFCCC entered into force less than two years after it was signed in Rio, and COP-1 took place one year later, in Berlin. A mandated COP-1 review of the adequacy of commitments produced three conclusions. First, most Annex I countries would not meet the UNFCCC's aim of lowering emissions by the year 2000—this could be predicted even in the absence of a specific target for reduction simply because emissions were expected to continue growing. Moreover, even if they did there were still two problems: there was no provision in the Convention for actions to be taken after the turn of the century, and the goal of stabilization of emissions at 1990 levels was in any case far too low to actually stem the problem of climate change. The parties therefore agreed to what came to be known as the "Berlin Mandate," the COP-1 agreement that set in motion the negotiation of the Kyoto Protocol, and an *ad hoc* Group on the Berlin Mandate (AGBM) was set up to carry this out (Breidenich et al. 1998, 318).

In making their decision to begin negotiations, delegates had before them a draft protocol that was actually introduced months before COP-1 by Trinidad and Tobago on behalf of a coalition formed in 1991, the Association of Small Island States (AOSIS). The membership of AOSIS comprises 43 small island and low-lying coastal countries that are very vulnerable to the threat of rising sea levels in particular,[24] and their proposal was correspondingly stringent. It called for Annex I parties to the protocol to reduce their CO_2 emissions by at least 20 percent from the 1990 baseline year by 2005, and to establish timetables for controlling emissions of other gases. It did not win much support from some other developing countries, given the huge divergence between the interests of AOSIS and those of developing countries such as OPEC members and China (Carpenter, Chasek, and Wise 1995b). The United States and Australia also stated objections to its inclusion of targets and timetables and its focus on CO_2 rather than on GHGs as a whole. Thus the final decision to begin negotiations contained, unsurprisingly, much weaker language than the original AOSIS proposal.

The issue of developing countries' responsibilities under new Protocol also arose again at COP-1, as a result of a German paper circulated earlier that tossed out the idea of placing commitments on developing countries according to their degree of industrialization. It was accepted by other OECD countries but consistently rejected by developing countries, perhaps because there was nothing offered in exchange for developing country commitments. In the end, the Berlin Mandate stated explicitly that the process toward a protocol would not introduce any new commitments for non-Annex I parties.

A major step forward was taken at COP-2. The lead US negotiator, Tim Wirth, stated that the United States was now willing to negotiate firm targets and deadlines for reducing GHG emissions.[25] One reason for this shift was the change in the US administration within a year after the UNFCCC was signed. The new President Clinton had chosen Al Gore as his vice-president, a self-avowed environmentalist who had already expressed particular concern about climate change. One result was a new interest in addressing the issue while attempting to mitigate the severe economic repercussions that could be expected from taking action to reduce emissions from the fossil fuel use upon which the US economy was so dependent. Thus, while holding out a possible acceptance of targets and timetables, the United States also insisted on "flexibility" in how targets could be met, a concept the United States had introduced at the second meeting of the AGBM in November 1995. The flexibility measures that the United States promoted were an outgrowth of its earlier calls for inclusion of all GHGs, attention to "sinks," such as forests, that absorb CO_2 emissions, and "joint implementation", which would give each country the right to claim credit against its own emissions target for financing projects in another country. Perhaps most significantly, such measures also included the concept of international "emissions trading", whereby countries or companies would be allocated credits, or "rights to emit" within an agreed international cap, and could buy more or sell unused credits to others depending on the relative costs to them of lowering their emissions.

As a result of this change in the United States' position, delegates at COP-2 were able to go beyond the Berlin Mandate, agreeing for the first time on a goal of negotiating legally binding targets. They set a deadline for doing this by COP-3, which would take place at the end of 1997. This was not an official

COP-2 decision, because several important countries objected, including the OPEC countries, Russia and Australia. The delegates who favored it were able to finesse this by including that goal in a "Geneva Ministerial Declaration." Despite the fact that this was not an official decision, the weight of the countries favoring this new goal, which now included the United States, was enough to turn the focus of the negotiations to its achievement, and this was indeed accomplished by December 1997 (Schröder 2001, 62).

There were four more meetings of the AGBM (eight altogether) before COP-3 in Kyoto, during which time three overarching thematic issues had to be negotiated: First and foremost, Annex I countries had to negotiate the actual targets for QELROs to which they would be willing to commit themselves. Second was that the agreement on targets would depend on simultaneous acceptance of the US-proposed "flexibility mechanisms", which raised many new questions, some of which were quite complex. Finally, there was the lingering question of new commitments by non-Annex I countries, which would bring with it the question of financial and technological assistance.

In the final months before Kyoto, the US Senate startled the world by raising a new "equity" objection to making concrete commitments to reducing GHG emissions. Earlier objections by the United States had alluded to the uncertainty of the science and the fear that the costs of reducing emissions would threaten the entire US economy, but the United States had finally acknowledged the overwhelming scientific consensus on human-induced climate change and on the measures needed to mitigate it and was developing ways to reduce the economic costs to itself of any reductions. The United States' new objection came in a US Senate resolution, proposed by Senators Robert Byrd and Chuck Hagel on July 25, 1997, which declared that "the US should not be a signatory to any protocol . . . which would (A) mandate new commitments to limit or reduce GHG emissions for the Annex I Parties, unless [it] also mandates new specific scheduled commitments to limit or reduce GHG emissions for Developing Country parties within the same compliance period, or (B) would result in serious harm to the economy of the US."[26]

This resolution threatened to make all earlier concessions and efforts moot. Democratic Senator John Kerry, who attended the Kyoto meeting as an observer, alluded there to the need for developing country commitments both in order to achieve the goal of mitigating climate change and in terms of equity, saying that he "couldn't see" that the Senate would "ratify an agreement . . . leav[ing] out developing country participation in the long run, because this is a global problem which cannot allow China, India, Brazil, Mexico, and other countries to believe they can just continue to pollute up to levels that [match] those of the United States or other people while we're busy trying to reduce."[27]

Senator Lott focused on the equity issue in even stronger terms in expressing his support for the resolution during the debate that preceded its adoption: "As written, this legally binding treaty would require the United States and other developed countries to reduce their carbon dioxide and greenhouse gas emissions to 1990 levels by the year 2010 . . . . And what would the developing nations contribute? . . . Nothing. . . . The treaty lets them off the hook. Mr. President, this is not an equitable international policy. This is not a level playing field for the United States. Unless all the citizens of the globe are involved, there is a clear inequity. . . . It means that we should only participate in a fair, balanced, equitable agreement where all nations must participate."[28] This equity argument was echoed in 2001 when the new President Bush announced that the United States would withdraw from the Kyoto Protocol, calling the treaty "unfair to the US" because it exempts developing countries from the first round of emissions cuts.[29]

But perhaps this equity argument was disingenuous. In Kyoto, Senator Kerry pointed to the main concern of the resolution's proponents: a Protocol that did not oblige developing countries to make the same commitments to GHG emissions reductions would be competitively disadvantageous to the United States.[30] In other words, it was again the economic cost to the United States of emissions reductions that was the real concern. Not only would the cost of reducing emissions increase the price of goods manufactured in the United States, but also the cost to the US economy would be even higher if the prices of products manufactured in the United States became uncompetitive in relation to similar ones produced in countries with no such regulations.

Might it be legitimate, though, to couch American fears of loss of competitiveness as an equity issue? Upon first glance, and without being informed of the history of the phenomenon of climate change and climate change negotiations, one might well think it unfair to the United States for a large group of developing countries—over 150—to make no commitment to emissions reductions during the first commitment period set out in the Kyoto Protocol while letting the United States bear what would undoubtedly be the biggest burden. However, recalling the causes of climate change and considering which countries have benefited more from the activities that produced the problem, one might accept as more equitable the norm that if one harms others one has a responsibility to accept the consequences and to ameliorate them. In torts law this can be translated into liability. If climate change is generally accepted to have been caused by the major industrial powers and if it holds the potential to bring harm to states which either are more vulnerable than the United States, such as small island states, or cannot afford to adapt to the effects of climate change, the principle of equity would, it appears, call for the opposite kind of response from the United States than that which Senator Lott advocate.

This norm was accepted by all countries, including the United States, when the UNFCCC was negotiated, through the language that acknowledged common but differentiated responsibilities and the notion that developed country parties should take the lead in combating climate change. In the Senate debate on its resolution in 1997, Senator Byrd acknowledged the argument that the United States was "responsible for the situation that has developed" and that therefore some would claim that "the US should bear the brunt of the burden". However, his only response was simply to say that "the time for pointing fingers is over".[31] Equity was not the real issue; if it were, the argument of equity for developing countries would easily trump any American concern over fairness to itself.

Nevertheless, even without the equity argument it is easy to see why all of the senators who voted on this resolution felt compelled to do so. During the Senate debate on the resolution a valid point was made that without (eventual) developing country commitments to mitigate climate change, there might be "no net reductions in global GHG emissions."[32] It was also realistic to predict that if "developing countries [were] left free to pollute the atmosphere" then they would "in so doing, siphon off American industries."[33] This remains a forceful argument in the United States. Yet there are still reasons to question this as the United States' position. First, the threatened movement of American industries overseas parallels the loss of American industries that happens through the process of globalization in any case if corporations move production to countries with fewer regulations and lower labor costs. Americans generally believe this is already happening,[34] yet it does not provoke the same level of hostility from American lawmakers in general.

Second, what about the apparent notion that developing countries were to be excused indefinitely from reducing their contributions to climate change? Was this ever the case? Despite newspapers and other sources that have latched on to the notion that the Protocol will "expire" after the year 2012,[35] subsequent commitment periods were always envisaged, which would include developing countries. The commitments in the Kyoto Protocol as it exists today were only agreed as the first step, again following the model of the Montreal Protocol. The fact that it is insufficient—there is no debate over this—does not necessarily doom it to failure, just as the Montreal Protocol as originally negotiated did not fail but has become more and more stringent with time. This despite giving developing countries various "grace periods" before requiring them to freeze, cut, and phase out their production and use of ozone-depleting substances. Negotiators for the United States understood this when they signed on to the Kyoto Protocol.

In addition, many developing countries have shown the willingness to take on more commitments, beginning with Argentina and Kazakhstan in 1998 (Cushman 1998, G4). Indeed, it is incorrect to think that developing countries do not already have obligations under the climate change regime as it has evolved so far. Even as long ago as 1992 they committed themselves to actions such as developing and publishing their "national inventories" of GHGs (UNFCCC Article 4.1(a)); formulating national programs with measures to mitigate climate change (UNFCCC Article 4.1(b)); promoting the development of technologies, practices, and processes that control anthropogenic emissions of GHGs (UNFCCC Article 4.1(c)); and promoting the conservation and enhancement of sinks and reservoirs of GHGs (UNFCCC Article 4.1(d)).

But this discussion leads to the question of how likely is it that developing countries are going to take concrete measures given the projected attitude of the United States and its rejection of concrete commitments to reduce its own GHG emissions? This leads back to the question of equity. What is truly equitable? In Kyoto, proposed protocol language on voluntary commitments for developing countries became entangled with arguments over emissions trading in the wee hours of the morning on the final day of negotiations. India, with China's support, took specific issue with the way that text on emissions trading had been negotiated—without participation by developing countries—and questioned how credits would be allocated initially. Should it be on the basis of egalitarianism, as per the argument that all humans "should be entitled to an equal share in the atmospheric commons"? Or should it be based on "historical responsibility", based on the idea that those who caused the problem should compensate those who suffer from it, bearing the primary burden for addressing it? A "willingness-to-pay" argument would suggest that those who are most concerned about solving the problem should bear the burden. Another argument could be made that, to be truly equitable, the scheme should be based on mitigating the international maldistribution of wealth. Alternatively, one could start from the position that countries have established some sort of legal right to emit at the level they are currently emitting, therefore any burden for reducing global emissions should be equally met by countries irrespective of their starting point in emissions.[36] This is the basis of the scheme that was finally agreed, and which prompted India and China's demands for the deletion of a draft article 9 on voluntary commitments for non-Annex I (developing) countries. It became a choice between seeing the negotiations fail altogether or removing that text, given the United States' insistence on emissions trading. The article was therefore removed.[37]

If the issue were really one of getting developing countries on board, why would the United States not take similar steps to those it was willing to take under the Montreal Protocol to ensure that this would happen? In that case, the United States accepted the need for an MLF in addition to the already-established GEF, partly as a result of complaints of bias and lack of transparency in the GEF as it was originally set up. In Kyoto, a number of developing countries called for a "clean development fund" (Bettelli et al. 1997). This concept transmogrified, however, into a "mechanism," the difference, as denoted, being in its purpose. The mechanism does not provide funding, but shall only "assist in arranging project funding as necessary." This difference was critical: in Kyoto, donor countries were unwilling to set up a fund along the lines of the MLF and were unwilling to open the door to that possibility in Protocol text. Clearly, it would have added to the costs of the Protocol for the United States to contribute to a new fund. Unlike in the case of the Montreal Protocol, the United States did not have enough interest in the Kyoto Protocol to take this measure to raise the value of agreement for developing countries.

On the other hand, the United States was able to lower the prospective costs of the Protocol to itself in Kyoto, through compromises that lowered its overall potential effectiveness. In taking the floor in the initial Senate debate after adoption of the Protocol, Senator Byrd acknowledged that the United States had achieved some victories in Kyoto. One "substantial victory", in his words, was the inclusion of free market mechanisms—emissions trading and joint implementation—in the text of the Protocol. These had been pushed hard by the United States. Their purpose was "to allow advanced nations and their industries to satisfy their requirement for emissions limitations by sharing, buying, and selling credits internationally, and to fulfill part of their obligations by assisting developing nations in developing cleaner technologies and conservation."[38] The final form of the Clean Development Mechanism also represented an American victory, in allowing US industries to earn credits for sharing pollution-reduction technologies and production processes. This would allow the United States to avoid some of the costs that it would otherwise have to bear in reducing GHG emissions domestically. Other major achievements by the American negotiating team in Kyoto were the inclusion of a provision allowing the purchase of emissions allowances from Russia, the inclusion of all six GHGs not covered by the Montreal Protocol, and a "generous" definition of carbon sinks.[39]

In the end, in the absence of agreement on an across-the-board formula for GHG emissions reductions, Annex I countries chose individual targets for their commitments. The United States' position going into Kyoto was to commit to a goal of reducing projected growth in GHG emissions by 30 percent by 2010, equivalent to stabilizing emissions at 1990 levels. In other words, this would achieve the same goal that many developed countries had wanted to include in the UNFCCC for the year 2000, but with a ten-year delay, so it was met with scorn by many delegations present in Kyoto. However, it seemed clear to many at the time that this was a negotiating position only.[40] Indeed, by the end of the conference the United States had agreed to a 7 percent reduction target: a 7 percent reduction in the three major GHGs from their 1990 levels, and a 7 percent reduction in emissions of the other three from their 1995 levels. Given the loopholes that the United States won, the actual burden this would place on the United States would be a 3 percent reduction of emissions the first commitment period of 2008–2012.[41]

This was not enough to allay US worries about the economics of climate change mitigation. During Senate debate on the Protocol in 1998, Senator Craig of Idaho cited a Wharton Econometrics Forecasting Associates prediction that compliance with the Kyoto Protocol would cost $250 billion, or 3.2 percent of GDP, and 2.5 million jobs.[42] Meanwhile, the benefits of mitigating climate change, or the risk of climate change, were much more difficult to calculate. With the worst effects not predicted to be felt until as much as 75 years later, the issue was replete with scientific uncertainty and associated discounting. In addition, the United States was not among the most vulnerable countries and, as the richest country in the world, might perceive itself as capable of undertaking adaptation measures if they should become necessary.

In the aftermath of Kyoto, therefore, the United States' position in climate change negotiations continued to be one of resistance. Because of the lack of encouragement from the one country that would have been the most capable of pushing through effective measures during the Kyoto negotiating process, the Protocol left many questions unanswered. A number of these were taken up in UNFCCC COPs between the Protocol's adoption and its entry into force, including the question of mechanisms for implementing the commitments made in Kyoto. Major disagreement between the United States and the EU regarding the flexibility mechanisms caused COP-6 to be suspended in 2000. Soon afterwards, the Bush administration came into office and "withdrew" from the Protocol.[43] The United States stuck to this position when COP-6 resumed in 2001, staying on the sidelines during discussions about the Protocol and making no new proposals.[44] In 2006, the United States' official position remained the same.

The Kyoto Protocol entered into force on February 16, 2005, over seven years after its adoption. As is common knowledge, its membership does not include the United States. But the new President Bush's withdrawal from the process in 2001 was not due to individual idiosyncrasy. There was actually little difference between this act and the fact that the Protocol had not been presented for ratification by President Clinton; indeed, President Bush's statement might have been counterproductive to his interests in that it drew attention to the issue, giving environmental activists a target for criticism. That there was no significant public backlash within the United States was compatible, however, with a calculation that the expected costs of the commitments to which US delegates had agreed in Kyoto continued to outweigh the benefits the United States could expect to receive.

US Interests and Global Climate Change

Throughout the era of international policy-making to address climate change, the United States has acted not as a leader but as what Sprinz and Vaahtoranta have termed a "dragger" (1994, 80). As a result, not only were its resources not available for use to manipulate other countries' interests to favor agreement, but other countries were pushed to try to adjust the value of agreement to win US acquiescence. This was done mainly by lowering the cost of agreement through downward compromises that lowered the effectiveness of the agreements made. This happened both in the UNFCCC, when the target date was removed, and in the Kyoto Protocol itself, with the incorporation of the flexibility mechanisms and other qualifiers that the United States insisted upon. Unfortunately, these were still not enough to offset the economic burden that the United States expected to incur. An examination of the breakdown of the United States' interests shows this clearly.

Environmental benefits: The environmental benefit from avoiding the potential effects of climate change are in many ways incalculable, given the uncertainties of the science behind predicted environmental losses. It is well accepted that there is some anthropogenic greenhouse effect (Lomborg 2001, 260). One early study, from 1991, estimated the economic damage to the US economy of human-induced climate change as only 0.25 percent of GDP.[45] However, another study, from 1992, argued that economic damage might reach $60 billion annually, or 1 percent of GDP, not including intangible costs such as loss of species.[46]

The science of climate change is beset with problems of predicting how, where, and to what extent the physical phenomena that make up the climate will alter with the changing composition of gases in the atmosphere. It is also impossible to separate completely the scientific and the political. The IPCC and other scientific institutions have long been subject to criticisms of Northern bias.[47] In addition, science itself, and the traditional caution of scientists in asserting certainty, has not been immune to exploitation by political leaders who appeal for more research into "scientific uncertainties" rather than into "discovering threats" (Boehmer-Christiansen 1994, 196; Soroos 2001, 5). Then there is the adversarial science mentioned in chapter 2, which results when competing interests on different sides of the issue fund research for the purpose of influencing policy to favor their respective positions. All of these vagaries make it possible for countries such as the United States to claim that the benefit of effective action is small, even while countries with an interest in advancing international regulation against climate change call the potential losses from ineffective action a potential "Katastrophe" (Benedick 1997).

Avoidance of economic costs: Unlike its situation during the ozone negotiations, the United States does not have in place domestic emissions standards that lessen its competitiveness vis-à-vis other countries with lower emissions standards. Rather, the United States finds itself in the opposite position, competing against countries with higher standards. Insofar as this gives the United States a "comparative advantage" for its exports, it has no interest in an effective agreement that would eliminate that advantage or in putting US industry in a less favorable position relative to exports from developing countries with no emissions reduction obligations.

Another economic argument for international cooperation, however, came out in a number of studies produced during the 1980s, which showed that reducing the inefficient use of emissions-producing fossil fuel could lead to economic gains through avoidance of certain costs. For example, a 1991 Report of the National Academy of Sciences concluded that "the US could reduce its GHG emissions by between 10 and 40 percent of the 1990 levels at low cost, or perhaps even with some net *savings*, if proper policies were implemented."[48]

The argument that the United States could avoid economic costs by reducing inefficient use of emissions-producing fossil fuel gained little traction, however. Why? Because its fuel prices are much lower than in many countries that must import fossil fuels and which therefore tax the use of fossil fuels, particularly petroleum, at a much higher rate. The United States is the world's largest producer of coal, oil, and gas (Oberthür and Ott 1999, 18). Even though it imports much of the oil it uses, the fact that it is a large producer of fossil fuels puts it in quite a different economic position from that of countries that must import most or all of these fuels. Given the fact that inefficient use of these fuels does not hurt the United States' balance of payments but, in fact, benefits some powerful economic actors within the United States, the United States has not had the same interest in reducing its reliance upon them. As a result, economic growth in the United States has not been "delinked" from growing use of fossil fuels to the extent that it has in some developed countries such as Germany and Japan (Rowlands 1995, 135). For these reasons many US policy-makers argue that policies to cut carbon dioxide will damage the US economy rather than enabling it to avoid economic costs.

Positive economic benefits: The decade that brought world attention to the threat of climate change began with the United States in an enviable position regarding research and development of energy-efficient and renewable-energy systems through federal funding and tax incentives. This could have put it in a competitive position to manufacture and market these products. It lost this position as the Reagan administration slashed funding and eliminated the tax credits[49] based upon newly mandated federal cost-benefit assessments and high discount rates[50] that gave more weight to short-term economic costs than to long-term economic gains. As a result, US industry has lost the opportunity for particularistic gains in many areas, as other countries, notably Germany and Japan, have outstripped the United States in development and marketing of alternative energies and energy-efficient systems (Rowlands 1995, 139, 143; Sprinz and Weiβ 2001, 82).

Costs of developing substitutes/halting activities: In 1990, the US Council of Economic Advisors estimated the costs to the US economy of proposals to cut the United States' carbon emissions by 20 percent by the year 2100 as ranging from $800 billion to $3.6 trillion (see endnote 42 above). The figures cited by Senator Craig regarding the cost of complying with the Kyoto Protocol, with far weaker commitments, was much lower, $250 billion. However, two 1999 studies came up with much higher figures than this. One study estimated the present value of the costs of implementing the Protocol as efficiently as possible as between $800 billion to $1,500 billion, compared to a present value of benefits of $120 billion; another study's cost estimate was in excess of $2.5 trillion.[51] The Department of Energy and Environmental Protection Agency came out with their own studies showing much lower overall cost estimates (see United States Department of Energy 1998; USEPA 2000). Thus far, however, the perception that the costs of compliance measures to the overall US economy has outweighted any benefits to the United States—even given the flexibility the United States itself built into the Protocol—has dominated American policy-making.

Costs of manipulation: The efforts the United States and other developed countries made to put developing countries' emissions on the agenda were emphatically rejected by most developing countries. The non-Annex I countries have repeatedly noted the failure of Annex I countries to live up to their commitment under the UNFCCC to take the first steps to combat climate change. Unless catastrophe should strike first that would raise the benefits to developing countries of taking action in their own self-interest, a demand that the G-77/China accept limits on their own GHG emissions will in all likelihood entail some *quid pro quo*. Any commitment to provide resources and new, cleaner technologies to the developing world appears, so far, to entail far more cost than the US in particular is willing to bear (Soroos 2001, 7).

Lomborg's notion of an "extended Kyoto Protocol" combines the goal of softening costs through emissions trading with the goal of achieving developing country participation through giving developing countries emissions permits that developed countries could then buy. This nevertheless still depends on resolving the issue of initial assignment of emissions rights, and would itself "involve a major redistribution from developed to developing countries" (Lomborg 2001, 305).

All in all, the United States has had good reason, in its own terms, not to undertake costly measures to mitigate climate change or to make commitments at the global level to do so, even in the light of international pressure. Although President Clinton signed the Kyoto Protocol a year after its adoption, upon Argentina's voluntary acceptance of a GHG emission target for the first budget period of 2008–2012 (Maathai and Trexler 1999), the changes in US administration—from George Bush to Bill Clinton to George W. Bush—were in the end mainly superficial. The basic cost-benefit equation as perceived by US policy-makers has not changed fundamentally since the problem rose to the international agenda in the 1980s.

CHAPTER 7

Conclusion: A Climate for Future Success?

Having now examined the history of global negotiations on ozone depletion, deforestation, and climate change, we are now in a position to consider the differences and similarities between the three cases and ask what the future may hold.

Comparing the cases according to the typology of potential benefits and costs for the United States that was set out in chapter 2, the reasons for the variation in outcomes becomes clear. There was an environmental benefit to be gained in all three cases, but the issue of ozone depletion was more salient to the US once evidence of an Arctic ozone hole was perceived as a threat to the Northeastern states. In terms of avoiding economic costs, because the United States had comparatively strict regulations that affected the producers of ozone-depleting substances such as CFCs (chlorofluorocarbons) and the timber industry, there was some interest in protecting those economic interests in the ozone and forest cases by pushing to "internationalize" US domestic legislation. The opposite was true in the case of climate change, because US production itself was lower in cost than production in other countries where the cost of fossil fuel-based energy was much higher. Those countries, such as Germany, became the pushers for international regulation to lower GHG (greenhouse gas) emissions, which would have the potential of raising the cost of production in the United States and leveling the playing field for their own industries.

Of the three environmental issues examined here, the ozone case presented a tremendous opportunity for the United States to accrue particularistic benefits, as international restrictions on the production and use of CFCs held great potential—ultimately realized—for helping American industries that were in an advantageous position in the development of CFC substitutes to garner windfall profits from marketing them.

In the case of both climate change and ozone depletion, halting activities that either produced the problem in question or were dependent on activities that did represented a prospective cost, as did the development of substitutes. The deforestation case differs in that the issue of shifting from economic activities that contribute to deforestation was never raised for the United States.

Finally, one of the most significant factors in all three cases was the potential cost of manipulation of other countries' preferences. In all three cases, the United States faced huge costs in attempting to induce developing countries to make commitments to alleviate the problem. In the ozone case it was willing to bear this burden; in the others it was not.

Table 7.1 below summarizes the similarities and differences in costs and benefits that an effective agreement in each of these areas would present to the United States.

Given the configuration of costs and benefits in the three cases, we can see why the similarities between the issues of climate change and deforestation are more significant than the fact that the ozone layer and the climate are both "atmospheric commons". This is true in terms of both the inadequate international responses to climate change and deforestation and the reasons behind their inadequacy. While the United States had a clear self-interest in leading the charge to develop effective international policies to address ozone depletion, this was not true for deforestation or for climate change.[1]

The three cases considered in this work have probably received more public attention in the United States than any other global environmental issues. If only one of the regimes examined in this work can be considered effective, in that it is on the path to achieving an actual amelioration of the problem it addresses, it is worth asking about the future prospects for creating effective regimes to grapple with the two failed cases.[2] If the predictions generated by my framework are correct, effective agreements will require a shift in US interests such that the perceived benefits of effective agreement will outweigh the costs.

In the case of deforestation, the will to proceed to an effective binding convention has diminished as public attention to this issue has declined and countries that had taken on leadership roles in this effort[3] seem to have given up, at least for the near term (Baldwin et al. 2005; Hajjar et al. 2006). For climate change, however, the situation is somewhat different as public attention to this issue is on the rise. This leaves the question of whether the United States' position on effective global agreement to stop the threat of climate change could ever shift.

Table 7.1 Summary of Cases

Cases			*Ozone depletion*[a]	*Deforestation*	*Climate change*
Outcomes[b]			*2*	*0*	*1*
Explanatory factors	*US leadership?*	at beginning negotiations	x[c]	x	o
	Willingness of lead state	at end of negotiations	x	(o)[d]	o
	benefits	environmental benefits	x (upward shift)	x	x
		avoidance of economic costs	x (upward shift)	x	o
		positive economic benefits	x (upward shift)	o	o
	costs	cost of substitutes	x	o	x
		cost of halting activities	x (downward shift)	o	x
		cost of manipulation	x (upward shift)	x (upward shift)	x (upward shift)

[a] Values here are based on a summary of values for four "sub-cases" on restrictions for different chemicals within the ozone case.
[b] 2= effective; 1 = ineffective; 0 = nonexistent (see chapter 3, table 3.1).
[c] x= factor present; o = factor not present.
[d] Although the US led the pro coalition, its unwillingness to provide adequate incentives or threats to change anti coalition opposition may be interpreted as an overall lack of will. Shortly after negotiations ended the United States officially changed its position to oppose a binding global treaty on deforestation.

The United States and the Future on Climate Change

A shift in position on international arrangements to address climate change will depend on a change in the United States' cost-benefit equation. This could happen with an upward shift, for any reason, in the amount of benefit to be gained from an effective agreement, and/or with a downward shift in the expected cost of agreement. Is this likely, or even realistic?

First, recall the unlikelihood that any pro coalition not including the United States could manipulate the value of the agreement or the no-agreement outcome to the United States, given the fact that the United States' own resources would not be available for this endeavor and the fact that its own wealth means it would perceive the amount of funding that could be offered as a carrot as even smaller. Any successful effort on the part of other countries to manipulate the United States' preferences to favor effective agreement would have to come from a powerful actor or actors who themselves favor effective agreement. This did happen once in the ozone case, when both the EU and China had an interest in pressing the United States to tighten restrictions on HCFCs (hydrochlorofluorocarbons) at MOP-11 of the Montreal Protocol in 1999 (see chapter 4). In the case of climate change, however, while the EU generally favors a more effective agreement than does the United States, China has so far not addressed the issue of US non-participation. Indeed, at the first conference of the parties acting as the first Meeting of the Parties to the Kyoto Protocol, it was the United States that courted China in hopes of swaying it to join the "Kyoto skeptics camp," not vice versa (Harvey 2006, 17). Furthermore, if such manipulation had to depend on raising the value of agreement for the United States, it is hard to imagine that the EU and China would ever be willing to offer a joint carrot.

Threats are costless if they work, so use of threats might be more feasible. However, a threat has to be credible to work, and, given that the only credible threats would probably concern trade restrictions, they would have to come from a state or states that had a large enough market to cause the United States a perceptible loss. Today they would also be subject to WTO prohibitions against unilateral trade measures—unlike the weaker rules that were in place when legislation was threatened in the United States in 1987 that would have blocked imports containing CFCs. Right after Bush's withdrawal from the Kyoto process in 2001, several European ministers gave veiled warnings that US relations with the rest of the world could suffer.[4] Nevertheless, no sanctions were ever attempted, an indication of how difficult it is for even relatively powerful countries to punish the United States, even in the face of near worldwide criticism.

It appears, then, that external manipulation of US interests is unlikely through either promises of rewards or threats of punishment. However, the possibility of coercion from other states is not the only, nor the strongest, influence on the United States' position. On the domestic side there are numerous factors that may be influential.

A vote loser? In 2002, Scott Barrett speculated that President Bush might come to see his stance against Kyoto as a "vote loser," given that he actually received fewer votes than Al Gore in the 2000 presidential race and only became president after, among other things, Gore lost votes to Ralph Nader, who was considered even "greener" than Gore.[5] In fact, the opposite happened, as Republicans gained congressional seats in 2002, Bush won reelection in 2004, and as of 2006 Bush had not changed his position on Kyoto. According to my framework, even if Gore had won it would probably have made little difference: rather than "withdrawing" its signature the United States would likely have maintained the status quo of nonratification.[6]

For an anti-Kyoto Protocol stance to become a "vote loser," it would have to be seen to be in opposition to a calculation of net benefits from the Kyoto Protocol for the United States. This will require a shift in at least some of the costs and benefits from Kyoto for the United States. It has not happened yet, but it may not be entirely outside the realm of possibility. The year 2005 may go down not only as the date that the Kyoto Protocol entered into force without the United States but also as the year that the United States' assessment of the costs and benefits of effective international cooperation on this issue began to shift. Using the typology of benefits and costs set out above, we can see how this could happen.

Prospective Shifts in Costs and Benefits to the United States

Environmental benefits: As time passes the threatened impacts of predicted climate change loom ever larger. Scientists have been warning for 20 years that we need to apply the Precautionary Principle and take action before the impact is felt, otherwise it may well be too late to halt climate change. One consolation for the lack of success in addressing climate change so far is that a presently seen problem can usually catalyze cooperative action more easily than future predictions can. The year 2005 saw an abundance of fresh scientific evidence that the future is here as regards climate change. The unusually high number of extreme weather events experienced in 2005 was seen by most scientists to be consistent with climate change and brought the largest financial losses ever experienced as a result of weather-related natural disasters, valued at more than US$200 billion.[7] The "first unequivocal link" between anthropogenic GHG emissions and temperature rises in the earth's oceans[8] was made in February 2005. This and related studies show a change in the circulation of waters in the North Atlantic Ocean, which scientists link to increasing amounts of fresh water in cooler regions due to polar ice melting and saltier water in the tropics and subtropics due to increased evaporation from rising temperatures. This phenomenon is described by the US National Academy of Sciences as likely to disrupt the Gulf Stream, which could lead to a significant cooling in Europe.[9] In April 2005 came news that ten of the hottest years on record have occurred since 1990[10]—up to that point the hottest decade on record had been the 1980s (Paterson and Grubb 1992, 294)—and that postulated agricultural benefits from global warming in some areas may not in fact be very likely. Studies were reported which showed the timing of hot temperatures can actually reduce the yield of crops if the hot weather coincides with their flowering.[11] In September, a report came out which showed that for the fourth consecutive summer sea ice floating on the Arctic Ocean had shrunk and was now at its smallest size in a century. This shrinkage has the potential to become irreversible, because dark open water increases heat storage, making it harder for ice to grow the following winter.[12]

Of greatest salience in the United States is global warming's potential effects on US citizens. In Alaska evidence grew that sea ice was retreating, causing erosion of the coast; permafrost was melting; more types of insects, including forest pests, were arriving; migration patterns of prey animals were changing; and the weather was becoming more unpredictable.[13] News of these phenomena captured the attention of Hillary Clinton and John McCain, who, along with a group of other US lawmakers, toured the affected areas in August of 2005 to publicize the problem.[14] Implications for human health, historically the factor that resonates most with the American public with regard to environmental issues, also began to be discussed. The potential effects of global warming on human health include not just the direct effects of higher temperatures but illnesses from viruses and other organisms long buried under deep ice in the Arctic and Antarctic regions that are now thawing.[15]

American attitudes regarding the issue of climate change were most noticeably shaken, however, by the catastrophe along the Gulf Coast at the end of August 2005. The study by Webster et al. (2005; see chapter 6) that was published just two weeks after Hurricane Katrina hit could not have come out in a more timely way for capturing the public's attention. Whether or not consensus can be reached on its conclusion that climate change is responsible for more extreme hurricanes, or even that more extreme hurricanes are taking place today than a generation ago, a link has been drawn between climate change and disaster in the minds of many Americans.[16] Indeed, concrete evidence of Katrina's effect in shifting attitudes regarding climate change was shown in the subsequent establishment of an "Evangelical Climate Initiative" by 86 evangelical Christian leaders. Acknowledging that many of their number had "required considerable convincing," the group announced in February 2006 that they had become "persuaded that climate change is a real problem and that it ought to matter to us as Christians." The group then linked images of Hurricane Katrina and global warming in television spots created to get their message across to their followers.

The most remarkable aspect of the Evangelical Climate Initiative is the fact that it relies upon the work of climate change scientists, even though evangelical Christians are associated with strongly held beliefs which in many instances have conflicted with scientific claims. Evangelical Christians do, however, form an extremely cohesive group and identify closely with strong leadership from within that group. The fact that some evangelical leaders are now modifying their own perceptions and attitudes strengthens tremendously the possibility of a shift among their followers, which could ultimately tip the balance of public opinion toward favoring international regulation of GHG emissions. Does this mean that group identity may matter more than economics in explaining any future shift in the official US position on climate change? No, because any shift in attitudes among evangelical Christians that stems from this Evangelical Initiative will have been the result of the shift in perceptions of the environmental costs of inaction that has now taken place among the leaders of that identity group.[17]

Avoidance of economic costs: Meanwhile, the value of other potential benefits from effective international agreement to address climate change may also be shifting upward. Barkin and DeSombre identify potential "positive feedback loops" that can encourage industry pressure for international regulation (2004, 2). In effect, this involves creating new costs from nonaction that industry then seeks to avoid, such as the costs created by the patchwork of state CFC regulations in the United States that led to an industry preference for national, and then international, CFC regulation in the ozone case.

There are several types of economic costs that may encourage industry preferences for effective regulation at the national or international level. These include, for example, costs to insurance companies, costs to companies wanting investment, costs of potential lawsuits, the costs of meeting differing regulatory standards in different areas, and the costs of competition from companies and countries not otherwise prepared to go as far toward reducing GHG emissions.

Details of a report commissioned by Ceres, a coalition of investors and environmental organizations, came out in late 2005 citing "a 15-fold increase in insured losses from catastrophic weather events" over the previous 30 years that had outpaced premium increases, inflation, and population growth during the same period. The report warns that insurance losses are likely to rise further in the next few decades, with "significant increases in hurricanes, floods, hailstorms, wildfires, droughts and heat waves" caused by rising global temperatures.[18] It notes that insurers are already reacting to increasing weather-related losses in Florida, Texas, and California through rate hikes and increasing restrictions on coverage, and are being pushed to do more to assess their "growing exposure from climate change." In an unrelated study, the Association of British Insurers calculated that, unless carbon emissions are reduced, the costs of Japanese typhoons could increase by two-thirds over the next 75 years, with the capital needed by insurers to cover severe storms rising by as much as $78 billion.[19] These reports are significant, as it is the insurance industry's job to identify and quantify catastrophic risks, in their own economic interest.

So far, however, the insurance industry has reacted "financially rather than politically" and has not pressed for regulation to slow climate change in order to decrease its exposure to growing risks of large payouts (Barkin and DeSombre 2004, 7). However, with Hurricanes Katrina, Rita, and Wilma occurring in rapid succession in 2005, followed by devastating tornadoes in Indiana, the US insurance industry may come to view this financial strategy as inadequate. One study reported in late 2005 noted that some insurance companies were already acting on their awareness of the risks of climate change in some ways, such as by reducing their own energy consumption to lower emissions of GHGs.[20] Another report stated that in the wake of Hurricane Katrina, the American International Group, the United States' largest insurer, was considering targeting investments toward companies involved in GHG mitigation.[21]

Targeting of investments is now occurring on a large scale and could grow into an area of huge potential costs. The Ceres group has again been influential in this. Composed of organizations and funds that together manage more than $2.7 trillion in assets,[22] the Ceres coalition was formed to press companies, regulators, and securities firms to address the long-term financial implications of global climate change.[23] In particular, the group has targeted the US Securities and Exchange Commission to push publicly traded companies to disclose the financial risks of global warming in their securities filings, to be analyzed as a matter of routine corporate financial disclosure.[24] In May of 2005, Ceres' Investor Network on Climate Risk, together with the United Nations Foundation and with the support of the UN Fund for International Partnerships, hosted the second Institutional Investor Summit on Climate Risk, which discussed climate risks and investor responses.[25]

With this much investment money at stake, there are signs that this investor movement is indeed having an effect on US businesses. The day before the Summit and apparently in response to the concerns it was about to address,[26] the head of General Electric (GE), Jeffrey Immelt, announced a new initiative by that company committing GE to reducing its GHG emissions by 1 percent by 2012 and the intensity of its GHG emissions—that is, GHG emissions per unit of income—by 30 percent by 2008.[27] While this falls far short of the commitment the United States would have had under Kyoto, it is still stronger than the voluntary target of decreasing emissions intensity by 18 percent by 2008 that the Bush administration touts. Moreover, Immelt took the opportunity presented by his announcement to call for government leadership and clarification of policy and made it clear that he would accept regulatory caps on emissions (see below). A *New York Times* editorial opined that Immelt "left no doubt that he believes mandatory controls on emissions . . . of CO_2 . . . are necessary and inevitable,"[28] although a GE spokesperson later clarified that GE was not "trying to make policy but more looking to push the agenda forward" given that "the US has not had a coherent energy policy for several administrations."[29] Mandatory controls can do what voluntary activity cannot: level the playing field between those companies that are willing to take voluntary steps to reduce emissions and those that are not,[30] thus enabling concerned and innovative businesses to avoid the economic costs of competition from foot-draggers. Mandatory caps on GHG emissions are what Kyoto is about; to admit the need for caps is a huge leap toward acceptance of Kyoto.

One reason for concern among some firms is the risk of lawsuits by their investors, for incurring higher costs by "unduly delaying emission reductions, damaging a company's reputation and failing to disclose investment-relevant information" on their risks from or contributions to climate change.[31] The desire to avoid potential costs of lawsuits from other quarters may also come to have some influence.[32] In 2004, a lawsuit was brought by eight states against several power companies to reduce CO_2 emissions from coal-fired power plants, and just after Bush's withdrawal from the Kyoto process in 2001, the question of whether poor countries that become victims of climate change could potentially sue GHG emitters such as the United States for compensation began to be discussed openly by a number of lawyers representing some of the largest environmental NGOs in the United States.[33] In 2002, the small Pacific island country of Tuvalu threatened to bring suit against the United States and Australia in the International Court of Justice. It remains to be seen how much of a legal threat will be posed from such quarters, however: the threat from Tuvalu has receded, thanks to New Zealand's agreement to accept all of its 11,600 citizens as they become displaced by rising sea levels,[34] and a judge meanwhile dismissed the US states' lawsuit against the power plants in September 2005, although an appeal was filed just before the end of that year.[35] There is a groundswell of interest in the possible legal ramifications of climate change for those most responsible for it, although many roadblocks exist for lawsuits.[36]

The cost to producers of varying regulations in different areas, as mentioned above with regard to the ozone case, also looms in the United States. A 2004 California regulation to lower vehicular GHG emissions by 29.8 percent by 2014 sets out the first global warming tailpipe standards in the world, according to one analyst.[37] While the big car manufacturers began legal challenges, six Northeastern states rapidly took up the idea of adopting similar standards.[38] In June 2005 the governor of California took his state a step further by declaring a target of reducing the state's GHG emissions to 2000 levels by 2010 and 1990 levels by 2020, and legislation to this effect was introduced in its State Assembly in April 2006.[39] This is 20 years later than the deadline for reaching the 1990 levels that most countries wanted to put into the UN Framework Convention on Climate Change in 1992 but, again, it goes much further than the US administration is willing to countenance.

Seven northeastern states also advanced a plan to regulate GHG emissions from power plants, through a program called the Regional Greenhouse Gas Initiative (RGGI) that would cap emissions and set up an emissions trading market.[40] The response of affected businesses was to resist, which caused four other potential member states to back away from the plan in December 2005, but if initiatives such as these withstand the legal and other challenges they face, this could be the beginning of a trend similar to that in the ozone case. In addition, the year 2005 saw some 195 US cities establish their own initiative to address climate change through an agreement to reduce GHG emissions by 7 percent. As of May 2005, municipal-level efforts to meet this target were already beginning to influence economic actors in some of these localities, reportedly in mostly positive ways.[41] Firms that do business overseas also face different standards, such as those of the European emissions trading scheme that came into effect in 2005 following on from the United States' demand for this type of flexibility mechanism under the Kyoto Protocol. The need to meet differing regulations puts numerous pressures on business, particularly with regard to calculating long-term investments. DuPont, for instance, was reported as being "out in front of the EU carbon trading program" in 2005, with close to $7 billion in annual sales in countries that have ratified the Kyoto Protocol.[42]

Finally, the argument that has most recently been used to justify the United States' refusal to ratify Kyoto—the lack of developing country commitments under the Protocol—might be turned on its head. Consider the spectre of rapidly increasing GHG emissions in developing countries such as China and India in the context of the commitment the United States and other developed countries made in the original UNFCCC to take the lead in addressing climate change. Is it realistic to think that many developing countries will commit to reducing their emissions if the United States does not do so first? Given developing countries' equity concerns and their demands for adequate funding to assist them with compliance, it would appear that a necessary condition for winning effective developing country commitments, and thus avoiding the economic costs of competition from countries without such commitments, may be the United States' own acceptance of Kyoto.

In fact, other costs may also be avoided by measures that could simultaneously address the need to lower GHG emissions. It is no secret that China is now the second-largest oil market in the world, after the United States, and that its increasing demand has contributed to high oil prices experienced in the US since 2005.[43] India, likewise, is rapidly increasing its oil imports (Bang and Froyn 2005, 14). Reducing American dependence on oil would lessen US competition with these developing countries over oil while carrying the potential to lower US fossil fuel emissions, including CO_2. It would also hold the additional benefit of lessening the security costs of the United States' dependence on Middle Eastern oil (over 60 percent of oil used in the United States is imported, most from the Middle East).[44]

Positive economic benefits: There is no question that European and Japanese innovators have gotten a head start in capturing the market for technologies to reduce GHG emissions or to replace GHG-emitting technologies. However, the GE initiative announced by Jeffrey Immelt in May 2005 was noteworthy in including a doubling of GE's investment in environmental technologies to $1.2 billion by 2010, a demonstration of Immelt's continuing belief that there is money still to be made in developing climate change-driven clean technologies. While noting that "Europe is the major force for environmental innovation," he pointed to a "vast new/profitable market in cleaner technology," saying that "America could create exports and jobs through making energy and environmental practices a national 'core competency'." Immelt aimed his remarks at getting the government on board with "shap[ing] technical priorities that create competitiveness in energy" and alluded to the helpfulness of "real targets, whether voluntary *or regulatory*" (author's emphasis added), "because they drive innovation."[44] Clearly, under this view mandatory caps would enable visionary American manufacturers to gain a positive economic benefit. Likewise, any move away from oil, such as oil-powered power plants, toward clean coal, could result in a positive benefit for coal industry actors.

While the coal industry has been a key influence on the current US administration's position against carbon dioxide emission caps, as the most affected domestic fossil fuel industry, Bang and Froyn see a possibility that this could change at some point in the future. Coal contains the highest amount of carbon per unit of useful energy, but if technological breakthroughs in the development of clean coal or carbon sequestration could lower or eliminate CO_2 emissions, it should be possible to split the coal industry away from the coalition that has long held sway in lobbying against policy measures to tackle climate change effectively (Bang and Froyn 2005, 15). Overall, the possibility that some important economic actors in the United States might, after a bit of investment, accrue positive economic benefits from regulation could potentially catalyze a shift in favor of effective agreement.

Costs of developing substitutes/halting activities: In 1999, the president of the United Mine Workers testified before a US Senate Committee that reducing CO_2 emissions in the United States in compliance with the Kyoto Protocol would cost a million jobs and over $100 billion in lost output per year.[46] Recall, however, that this was just two-fifths of the cost to GDP and to jobs that US Senator Larry Craig had put on Kyoto compliance just one year earlier. It was also a far cry from a return to horses and candles that was predicted when fears of climate change came to the fore in the early 1980s. Fortunately it has indeed been possible to develop substitutes for many of the substances and activities that produce GHG emissions. For example, a reduction in emissions does not have to depend on reducing vehicle use if the effort to lower fossil fuel use carries the "second-order effect" of advancing the development of alternative climate-friendly fuels (Bang and Froyn 2005, 1). As cleaner technologies are developed, the costs of compliance come down and the spectre of severe negative repercussions for the US economy is reduced.

As for the costs of developing the substitute themselves, these are offset if they produce positive economic benefits. They may therefore be rightfully regarded as an investment by people such as Jeffrey Immelt. Moreover, government subsidization of development of substitutes since the 1970s lowers the cost of substitutes, at least slightly, and these subsidies are increasing.[47] By lowering the cost of substitutes, subsidies also lower the cost of international regulation, and may therefore contribute to reducing opposition to it among actors who benefit from them. Such incentives may have contributed to the shift in the positions of many oil and car industry actors that led to their gradual abandonment of the Global Climate Coalition (GCC), an industry alliance that had lobbied strongly against international regulation since before the negotiation of the Kyoto Protocol but ultimately collapsed in 2002. Many former GCC members now show a readiness to accept "the need to convert our carbon-based energy economy into a hydrogen-based energy economy."[48]

Cost of manipulation: The cost of manipulating other countries' interests may in the end mean that the costs of a truly effective international agreement will continue to outweigh the benefits. There are two reasons, however, why this might not forever be so. First, China has begun to perceive costs from emissions caused by its use of coal, in the form of air pollution as well as the threat of climate change.[49] This means that China's preferences are shifting in the direction of more effective agreement anyway. The effect of this should be to lower the cost of manipulation for the United States; particularly if other countries' perceptions of costs likewise shift. Second, as seen from the above discussion, the benefits of effective agreement may eventually come to outweigh the costs in other areas, thus lowering the hurdles to overcome in the effort to achieve US compliance with international emissions limits. If there were net gains to be made from international climate change regulation otherwise, it might then be in the United States' interest to take on the cost of manipulation, particularly if that cost itself were to diminish.

The possibility that a trend may be developing within the United States toward greater acceptance of international emissions caps in general or the Kyoto Protocol in particular has been noticed by both the media and academics. Of particular note is that in June 2005 the US Senate "superseded" its own 1997 resolution opposing ratification of the Kyoto Protocol with a resolution endorsing "mandatory, market-based limits and incentives on emissions of greenhouse gases" in the future.[50] Even George W. Bush, long associated with skepticism over whether global warming exists and whether it can be linked to human causes,[51] agreed to a statement issued by the leaders gathered at the G-8 Summit in July 2005, that "increased need and use of energy from fossil fuels, and other human activities, contribute in large part to increases in greenhouse gases associated with the warming of our Earth's surface."[52] This came shortly after the chief of staff of the White House Council on Environmental Quality resigned after being shown to have watered down language in government scientific documents on climate change in ways that favored the oil industry.[53]

On the other hand, this watering down of scientific language was only one of several actions that were arguably part of a broader administration-wide effort to deceive the American public about the true environmental costs of climate change. This deception also included allegedly suppressing research findings that indicated possible connections between climate change and hurricanes, even before Katrina, as well as other evidence of global warming.[54]

In light of such attempts to censor information on the costs of climate change, one might question whether rational calculations of costs and benefits can really explain US attitudes in the climate change case. Rational decision-making requires full information, which such efforts to suppress scientific evidence are intended to prevent. Arguably, however, these acts of censorship may have had the opposite effect than what was intended, given that the scandal thus generated has received much more media attention than the mere presentation of the evidence itself would have garnered. In hindsight, the endeavor comes across as a failed effort to swim against the tide of evidence of the real costs and benefits of international cooperation on climate change. It may therefore be likened to Anne Gorsuch's ultimately failed attempts to stem the tide toward international CFC regulation in the 1980s, as described in chapter 3.

However, there are also other factors that may impede progress toward an effective international climate regime based on full US participation. A major consideration is whether, even with full acceptance of the scientific predictions, US actors might perceive adaptation to climate change as more cost-effective than mitigation. Indeed, in 2005 a joint IPCC working group began to consider the need to develop "alternative funding paradigms" for local adaptation in developing countries, given the GEF's focus on global environmental benefits and the fact that, as an IPCC vice-chair put it, "we've realized how difficult it is to cut emissions."[55]

In addition, an Asia-Pacific Partnership on Clean Development and Climate, negotiated in secret between six countries, including the United States, stirred suspicion among NGOs when it was announced in late July. This agreement offered a model emphasizing voluntary technological partnerships, expansion of markets for renewable energy and distributed generation, and voluntary emissions reductions goals based on emissions intensity. There was much speculation, however, that it was intended as an alternative to Kyoto.[56]

Britain's Prime Minister, Tony Blair, seemed to move toward acceptance of this as the way forward at a "Global Initiative" meeting hosted by former President Clinton in New York in September of that year.[57] Even more significantly, in January 2006 Canada's new Conservative government was reported to be considering "what it would mean for Canada to become the seventh country in the Asia-Pacific partnership on climate change.[58]

And what of the Kyoto process in the absence of the United States? At a meeting of an *Ad Hoc* Working Group on Further Commitments for Annex I parties under the Kyoto Protocol (AWG) in June 2006, delegates agreed to begin negotiating further commitments for Annex I countries during the post-2012 phase. This is, however, only one track of a two-track process that was agreed at the joint COP/MOP meeting of the FCCC and the Kyoto Protocol in December 2005. The United States managed to get an agreement at that time on continuing a separate "non-binding dialogue" under the UNFCCC itself (Decision FCCC/CP/2005/L.4/Rev.1).[59]

At this time the jury is still out on whether the United States will eventually come into the Protocol, given the commitments that would entail, or continue to try various alternative paths, which at least give the appearance of trying to do something even though they do not entail the costs of real commitments. Meanwhile, although other countries are proceeding ahead with the Protocol, it was reported that leaked instructions to the Canadian delegation to the *Ad Hoc* meeting in June 2006 specified that any new agreement "must include the USA and all major developing country emitters," thus lending support to my argument that major new commitments on GHG emissions reductions from other countries are unlikely without similar commitments from the United States.[60]

Other Factors

Are there other factors that could influence whether the United States' position on climate change is likely to shift? Specifically, are there factors other than perceptions of costs and benefits at the national level that may help to account for differences between the US and those countries that have been the strongest proponents of international regulation of GHG emissions?

Europeans have generally been more concerned about global warming and other environmental problems than the United States, but this can be largely attributed to a different cost-benefit calculation for the countries that have taken the lead on environmental issues. In Germany, three factors are different. First, rather than having a large and economically powerful fossil fuel industry as the United States does, Germany must import most of its energy. Second, Germany, among other European countries, has felt the effects of several perceptible environmental problems, all with negative economic consequences: acid rain, the after-effects of Chernobyl (Weidner 2002, 154, 155), and the effects of global warming itself, which is now causing snow to melt in the Alps.[61] Finally, and partially as a result of these factors, Germany has taken a strong lead in developing cleaner energies and cleaner technologies in general, so it stands to profit from encouraging other countries to reduce emissions (Weidner 2002, 190).

There are two other factors, however, which do not fit as well into a pure cost-benefit analysis. First, the parliamentary systems of most Western European countries, including Germany, differ from the presidential system of the United States. The fact that most are based on proportional representation has made it possible for Green Parties to become participants in making policy rather than simply lobbying policy-makers from outside government.

This difference has been credited with making Germany favor more "pro-environmental" policy than would otherwise have been the case (Vig and Faure 2004, 7). However, this would not explain the greater concern for global warming on the part of Britain's prime minister, Tony Blair, as British Members of Parliament are elected to office through a single-member-district system more similar to that of the United States and the British Greens have no seat in that Parliament.

Constructivists would argue that there may be something else underlying differences between European pro-Kyoto countries and the United States: a difference in perceptions of interests that differences in obvious costs and benefits cannot fully explain. Perhaps average discount rates vary among countries, for instance. Why should this be? Constructivists assert that differences in identity groups influence people's interests. If neo-Gramscian views on the hegemony of ideas are correct, the ideology of individualism and market exchange that prevails among those in power in the United States may influence the way in which most Americans perceive their interests in a profound way. Longer-term interests for the society as a whole may therefore be less salient for most Americans than they are in European countries that typically have more socially oriented policies on their books. Americans may also be susceptible to an "issue-attention cycle", in which individual preferences for effective action on an environmental problem may depend more on how dramatically it is portrayed in the media, as well as how costly it turns out to be and how quickly other issues come to replace it in prominence, than on more objective individual cost-benefit calculations (Downs 1972); it is unclear whether this theory applies across countries.

Differences between the United States and Europe would only matter for the issues discussed here, however, if Europe, in the form of a united, sovereign state, were to become the dominant power of the international system. In such a case, costs and benefits as perceived by Europeans would matter more than American assessments. Similarly, when China becomes dominant, assuming it maintains its current trajectory, Chinese perceptions of China's interests will matter most. For the time being, however, as long as the United States is the international powerhouse, the characteristics of "American-ness" are the factors of importance to the creation and development of international environmental regimes. It is an American perception of costs and benefits that determines the interest of the United States, and that, so far, determines the effectiveness of international environmental cooperation.

Notes

Chapter 1 An Introduction to Three Global Environmental Issues

1. United Nations Food and Agriculture Organization. *Global Forest Resource Assessment 2005*. Web address: http://www.fao.org/forestry/site/fra2005/site/en.
2. See chapter 6 for a list of these gases.
3. US Department of Energy, *Mitigating Greenhouse Gas Emissions: Voluntary Reporting*. Energy Information Administration, Office of Integrated Analysis and Forecasting, US Department of Energy, Washington, DC, October 1997.
4. Research findings published in early 2006 indicate that ozone recovery detected in the 1990s may have been at least partially attributable to the intensity of the eleven-year solar cycle. However, although the solar cycle may complicate attempts to determine the effect of decreases in ozone-depleting chemicals, scientists predict that a sustainable ozone layer recovery should be detectable after the sun goes through its next period of minimum intensity in 2008, and may still recover by 2050, as per earlier UNEP predictions. David Adam, "Hole in ozone layer expected to increase," *Guardian*, February 16, 2006, 13; also see Naila Moreira, "Ozone 'Recovery' May be Solar Trick," *ScienceNOW Daily News*, February 13, 2006. Web address: http://sciencenow.sciencemag.org/cgi/content/full/2006/213/1?eaf.
5. See, e.g., Steve Connor, "The Final Proof: Global Warming is a Man-Made Disaster," *Independent*, February 19, 2005, 1; United Nations Framework Convention on Climate Change, "Current Evidence of Climate Change." Web address: http://unfccc.int/essential_background/feeling_the_heat/items/2904. php.
6. US official, personal communication, August 2001; British official, personal communication, May 2005.

Chapter 2 A Different Approach to Understanding Environmental Regime Creation

1. My approach is based on the negotiation analysis approach of Sebenius (1991, 1992a, 1992b; Lax and Sebenius 1986) and others and Knight's (1992)

power-based theory of institution building in the context of conflicting interests. See also Krasner's third definition of bargaining power as the ability to change the values within the game's payoff matrix (1991, 340).

2. S. Touval and J. Rubin, "Multilateral Negotiation: An Analytic Approach," Cambridge, MA: Working Paper Series, Program on Negotiation, Harvard Law School, 1987, 1; quoted in Zartman 1994. The same observation can be made about "dilemmas of common aversions" that Stein models with coordination games (1983, 126); also see R. Hardin 1982.
3. The authors acknowledge that the bargaining process itself is a potential source of change in preferences but, justifiably enough, exclude it from the scope of their article.
4. Parties to an agreement include those states that have ratified it or undertaken an equivalent procedure, including acceptances, accessions, and approvals for those states requiring them; in many states this necessitates legislative approval of the executive's signature on a treaty.
5. A shorter version of this argument appears in Davenport 2005.
6. W. M. Habeeb, *Power and Tactics in International Negotiation* (Baltimore, MD: Johns Hopkins University Press, 1988); quoted in Zartman 1991, 68.
7. Snidal (1985b) is optimistic about the chances for successful and stable cooperation among a small group of smaller powers in the absence of a country with overwhelming power superiority. However, his analysis does not address a situation in which a coalition of smaller countries would be required to cooperate in order to coerce cooperation by a dominant power that opposes agreement; the likelihood of this appears very small.
8. Lisa Martin, *Coercive Cooperation: Explaining Multilateral Economic Sanctions* (Princeton: Princeton University Press, 1992); cited in DeSombre 2000 Perceptions may also be affected by the framing effect identified by Tversky and Kahneman (1986; also see Berejekian 1997), or the phenomenon found in numerous studies that costs are generally felt more than benefits. Thus, even if everything else were equal, the costs of manipulating another state's preferences through incentives might be felt more deeply by the manipulating state than would the benefit be felt by the target state. This might influence to some small extent which form of manipulation is chosen, but it appears that the feasibility of each type of manipulation in specific circumstances has much more to do with which one is chosen than a framing effect.
9. US official, personal communication, December 2001.
10. US official, personal communication, December 2001.
11. See Haas 1992a; also see, e.g., Benedick 1998.
12. Snidal (1985a) is incorrect in calling environmental services "public goods"; see Hardin 1992, 17, fn 4.
13. National interests are not monolithic, of course. Nevertheless, the unitary actor approach may be used in considering policy-makers' calculations of net costs and benefits of agreement in overall terms. As Snidal points out, "domestic factors shape the preferences that guide states in their interactions. But if the executive has maintained its policy (or implemented coherent policy change) then the unitary actor assumption is sustained" (Snidal 1990, 340–342).

14. Oye and Maxwell contrast "Stiglerian" situations with "Olsonian" ones in which the converse is true (terms based, respectively, on George Stigler, "The Economic Theory of Regulation," *Bell Journal of Economics* 2 (1971): 3–21, and Mancur Olson, *The Logic of Collective Action: Public Goods and the Theory of Groups* (Cambridge, MA: Harvard University Press, 1965).
15. Australian NGO representative, personal communication, August 2001.
16. See, e.g., Sebenius 1992b, 335.
17. Taken from Nicole T. Carter, CRS Report for Congress. "New Orleans Levees and Floodwalls: Hurricane Damage Protection," Congressional Research Service Report RS 22238, September 6, 2005. Also see Will Bunch, "Did New Orleans Catastrophe Have to Happen? 'Times-Picayune' Had Repeatedly Raised Federal Spending Issues,"*Philadelphia Daily News*, August 31, 2005. Web address: http://www.editorandpublisher.com.
18. Delegate to International Tropical Timber Council meeting, personal communication, May 1994.

Chapter 3 Ozone Politics

1. Stratospheric ozone is distinguished from ground-level ozone that is itself anthropogenic in origin and is one of the chemicals that contribute to smog.
2. Richard Benedick was the chief US negotiator for the ozone treaties from 1985 till the mid-1990s.
3. Cited in Thomas 1992, 201.
4. See Appendix 4.1 for a detailed list of commitments on ODSs, with targets and timelines.
5. For CFCs as well as for all other chemical groups listed, some exception is made for essential uses.
6. Except where otherwise noted, article numbers refer to articles in the Montreal Protocol.
7. According to Benedick, because many early compliance problems, particularly for developing countries, were due to the great technical complexities in assembling national data on production and trade that was the foundation for monitoring compliance, there was a good rationale for providing an option for "friendly" assistance rather than stigmatization through cautions (1998, 272).
8. Trade sanctions were originally intended to encourage countries to join the Protocol by helping to ensure that markets for CFCs would dry up for non-parties as more and more countries acceded.
9. R. E. Train, EPA, speech to NATO Committee on the Challenges of Modern Society, December 3, 1976; cited in Parson 2003, 45, fn 102. Even as early as May 1975 the United States and Canada asked the OECD, which had the authority to negotiate agreements among its member states, to take action on CFCs, but after publishing a staff report examining an aerosol ban in 1980, OECD activity ground to a halt (Parson 2003, 45).
10. Benedick 1998, 41. As of 1990, the United States supplied about 30% of world demand for CFCs, second to the ECs 40% (Jachtenfuchs 1990, 261).

11. G. W. Wirth, P. W. Brunner, and F. S. Bishop, "Regulatory Action," *Stratospheric Ozone and Man* 2(1981); quoted in Morrisette 1989, 806.
12. Adele R. Palmer et al., *Economic Implications of Regulating Chlorofluorocarbon Emissions from Nonaerosol Applications* (Santa Monica, CA: RAND Corp., 1980); cited in Brown and Lyon 1992, 128. Also see Rowlands 1995, 102.
13. DeSombre (2000, 27–28) is correct that US industry later had an incentive to push for international regulation regarding CFCs in non-aerosol uses, but not until 1986.
14. Reported in Parson 2003, 115.
15. Mostafa Tolba was Executive Director of UNEP from 1976 to 1992.
16. Parson in fact calls the period between 1975 and 1985 a "decade of deadlock" (2003, 245). I argue that the deadlock persisted until negotiations leading to the Montreal Protocol began to produce convergence in positions in 1987.
17. Nigel Haigh, EEC *Environmental Policy and Britain*, 2nd ed. (London: Longman, 1989), 266; quoted in Benedick 1998, 25.
18. Parson 1993, 32; Grundmann 1998, 206; see also Brown and Lyon 1992, 125; DeSombre 2000, 145.
19. US Environmental Protection Agency, "Stratospheric Ozone Protection Plan." *Fed. Reg.* 51 (1985) 1257; cited in Benedick 1998, 49, fn 22.
20. Dudek and Oppenheimer (1986) summarize the findings of a number of these studies.
21. *Congressional Record*, S7712, June 5, 1987.
22. Testimony of Eileen Claussen (Director, EPA Office of Program Development); and Lee Thomas (EPA Administrator) in *Joint Hearing Before the Subcommittees on Hazardous Wastes and Toxic Substances and on Environmental Protection of the Senate Committee on Environment and Public Works*, 100th Congress, 2nd Session (1988) at 283, 333, 472, and 525; cited in Shimberg 1991, 2187.
23. ARCFCP Policy Statement, September 16, 1986. In Alliance for Responsible CFC Policy 1987, I-3.
24. See, e.g., Haas 1992a, Benedick 1998, and Porter, Brown and Chasek 2000.
25. 132 *Congressional Record,* S15, 678–79 (October 8, 1986). Senator Chafee did not actually introduce this legislation until February 19, 1987.
26. *Congressional Record* S2289 (February 19, 1987); quoted in Shimberg, 2187; Rowlands 1995, 114.
27. Interview with Tony Vogelsburg of DuPont; cited in Litfin 1994.
28. DuPont letter to customer, September 27, 1986; available in Congressional Hearings, Senate Committee on Environment and Public Works, January 28, 1987, at 172–175; cited in Baldwin 1999, 114.
29. Malcolm Gladwell, "Du Pont Plans to Make CFC Alternative," *Washington Post,* September 30, 1988.
30. By April 1987, DuPont had in fact patented some of its safer alternatives (James Erlichman, "Britain Blamed as Ozone Talks Face Breakdown," *Guardian*, April 29, 1987).
31. "DuPont Position Statement on the Chlorofluorocarbon/Ozone/Greenhouse Issues," *Environmental Conservation* 13:4 (1986): 363–364.

32. Interview with Joseph Glas, cited in Litfin 1994, 94; also see Doniger 1988; Rowlands 1995, 114.
33. *Chemical and Engineering News*, November 24, 1986, 49; cited in Parson 2003, 123.
34. Sprinz and Vaahtoranta 1994, 93; also see "Why Toy With the Ozone Shield?" *New York Times*, December 16, 1986, A34.
35. See, e.g., Alan MacGregor, "Deal Near on Saving World Ozone Layer," *Times* (London), April 30, 1987, 10.
36. See, e.g., Thomas W. Netter, "U.N. Parley Agrees to Protect Ozone," *New York Times*, May 1, 1987, A1; see also Roan 1989.
37. Litfin 1995, 262; Doniger 1988, 89–90; also see Roan 1989.
38. The original legislation calling for a 95% cut was ultimately added in modified form to the 1990 Amendments of the Clean Air Act, after international regulations were similarly tightened (Shimberg 1991, 2179–2180).
39. Nigel Haigh, in correspondence with Richard Benedick, August 3, 1988; quoted in Benedick 1998. See also Bellany 1997, 149–150.
40. Jim Losey, EPA; quoted in Roan 1989, 210.
41. James Erlichman, "Ozone Layer Hangs on British Thread," *Guardian* (London), May 1, 1987.
42. Statement of Senator John H. Chafee, in 132 *Congressional Record*, S14, 678–679, October 8, 1986. Also see Rowlands 1995, 115; DeSombre 2000, 146; Shimberg 1991, 2186.
43. DeSombre 2000, 237; see also 132 *Congressional Record*, S15679, October 8, 1986.
44. "Our Planet is Losing Its Ozone Layer," *Le Soir*, February 3, 1987; quoted in Shimberg 1991, 2178; also see Parson 2003, 131.
45. Congressional testimony (written), A.D. Bourland, February 8, 1990; cited in Parson 2003, 199, commenting on a subsequent proposal for legislation, HR 2699.
46. Quoted in "Environment: US Criticizes Europe Over Protection of Ozone," IPS-Inter Press Service/Global Information Network, February 27, 1987.
47. James Erlichman, "Pushing Britain in to a Cheap Solution: Why We Are Ready to Help Stop Ozone Pollution," *Guardian*, September 4, 1987.
48. See Benedick 1998, 81; Parson 2003, 145.
49. "Hole in Arctic Ozone is Feared," *New York Times*, May 18, 1988, A25; Rowlands 1995.
50. See Benedick 1998, 121; Parson 2003, 161.
51. "Du Pont Sends a Message on Ozone," *New York Times*, March 29, 1988, A26; Malcolm Gladwell, "Du Pont Plans to Make CFC Alternative," *Washington Post*, September 30, 1988, F5; "Du Pont Produces CFC Replacement on Small Scale," *Journal of Commerce*, February 3, 1988, 9B; Rowlands 1995, 118; Parson 2003, 156.
52. Joe Steed of DuPont in an interview with Litfin; quoted in Litfin 1994, 126.
53. Natural Resources Defense Council, "Lawsuit Seeks Full US Phase-out of Ozone-Depleting Chemicals," *NRDC Newsline* (November/December 1988): 4; cited in Litfin 1994, 127.

54. Litfin 1994, 128, quoting Kevin Fay of the ARCFCP and Stephen Seidel of the EPA, respectively.
55. "Chlorofluorocarbons and their Effect on Stratospheric Ozone," Department of the Environment, Central Unit on Environmental Pollution, Pollution Paper No. 5 (1976); "Chlorofluorocarbons and their Effect on Stratospheric Ozone" (Second Report of the Department of the Environment), Pollution Paper No. 15 (1979); ICI Mond Division, "Chlorofluorocarbons and the Ozone Layer: An Appraisal of the Science," October 1986; and U.K. Stratospheric Ozone Review Group, "Stratospheric Ozone," August 1987; cited in Maxwell and Weiner 1993, 21, 22, and 28 respectively.
56. Indian representative to the Montreal negotiations in 1987; quoted in Benedick 1998, 100–101.
57. Quoted in Rowlands 1995, 170.
58. These were Egypt, Ghana, Kenya, Mexico (the only producer country), Panama, Senegal, Togo, and Venezuela (Rowlands 1995, 185, fn 6).
59. These were Egypt, Kenya, Malta, Mexico, Nigeria, Singapore, Uganda, and Venezuela (Rowlands 1995, 185, fn 12).
60. Ziul Rahman Ansari, Indian Minister for Environment and Forests; quoted in Richard North, "Appeal for Fund to Help Third World Cut CFCs," *Independent*, March 8, 1989, 6.
61. "History of the Montreal Protocol's Ozone Fund," *International Environmental Reporter*, November 20, 1991, 636–640.
62. "US to Join Fund to Help Curb Ozone Depletion," *Los Angeles Times*, June 16, 1990, A27.
63. Ibid.
64. Rowlands 1995, 185, fn 16.
65. UNEP official, personal communication, August 2005.

Chapter 4 Ozone Protection: The Story Continues

1. Depledge et al. 2001; Dan Bilefsky, "ICI Issues Warning on Climate Change Proposals," *Financial Times*, May 13, 2000, 4; Brown and Lyon 1992, 129; Parson 2003.
2. Today in developed countries, where CFCs have been phased out altogether, HCFC use represents 13% of all former CFC uses, according to the Alliance for Responsible Atmospheric Policy (formerly ARCFCP). Web address: http://www.arap.org/docs/hcfc-hfc.html.
3. Friends of the Earth, *Hold the Applause!* (Washington, DC, 1991), 43; quoted in Litfin 1994, 125. The official demarcation of CFCs regulated under the Protocol is "fully halogenated CFCs" (UNEP 1983).
4. Forest Reinhardt, "DuPont FREON Products Division: Prepared as a Harvard Business School Case" (Washington, DC: National Wildlife Federation, 1989); cited in Litfin 1994, 125.

5. Tony Vogelsburg of DuPont; quoted in Litfin 1994, 150.
6. ICI, "HCFCs—The Low ODP Solution," in "The Ozone Issue and Regulation," brochure, Runcorn, Cheshire, June 1990; quoted in Benedick 1998, 137.
7. Report of Mop-2 (UNEP/OzL.Pro2/3); Annex VII, section II; quoted in Benedick 1998, 175.
8. *Economist*, November 28, 1992, 50; quoted in Parson 2003, 219.
9. According to Parson (2003), patented HCFC blends promised prices as much as 10 times higher than those of CFCs.
10. At MOP-11 in 1999, Switzerland and Greenpeace also voiced concerns about industry groups' over-representation on the TEAP, resulting in a bias in its conclusions with regard to HFCs and PFCs—both chemicals that could fulfill many of the uses to which HCFCs have been put (Depledge et al. 1999, 13).
11. See "CEO's to US: Oppose Accelerated HCFC Phase-Out," *Engineered Systems* 14:7 (July 1997): 20.
12. NGO observer at Beijing, personal communication, December 2001.
13. Ibid.
14. Ibid.
15. US manufacturers are now beginning to move away from HCFCs (RAND 2006), at least partly thanks to the fact that production of HCFCs is now frozen in developed countries as of 2004.
16. See web address: http://www.suva.dupont.ca/.
17. Former EPA official, personal communication, July 2005.
18. Ibid.
19. Tom Donahue, atmospheric scientist, University of Michigan; quoted in Dotto and Schiff 1978, 93.
20. Richard Starnes, "Scientists say UN Bureaucrats Rewrote Findings," *Ottawa Citizen*, August 28, 1997, A1.
21. UNEP 1997. Corrigendum to the April 1997 TEAP Report. Web address: http://www.unep.org/ozone/teap/Reports/TEAP_Reports/Crigndm.pdf.
22. Don Amerman, "Us Producers Seek to Peel Away Barriers; Phytosanitary Bans Still Prove Biggest Hurdle; Smaller Foreign Crops will Help Sales," *Journal of Commerce*, October 17, 1997, 7A; Robert Steyer, "Growers Brace for Jolt From Fumigant Ban,"*St. Louis Post-Dispatch*, May 5, 1997, 12; Peter Fairley, "Clinton Pressured To Ease 2001 Ban," *Chemical Week*, October 16, 1996, 15; Parson 2003, 218.
23. In 1999, funding for the replenishment of the MLF actually decreased, from $466 million during the 1997–1999 triennium to $440 million for 2000–2002. This discrepancy is somewhat offset by the fact that many countries have not ratified the Amendments requiring controls on MB (Oberthür 2000, 40). Ratification by China in particular will have implications for levels of MLF funding, a need that donors will have to address.
24. US EPA, 40 CFR Part 82, "Protection of Stratospheric Ozone: Incorporation of Clean Air Act Amendments for Reductions in Class I, Group VI Controlled Substances," Federal Register: November 28, 2000 (Volume 65, Number 229).

25. Jerry Naunheim Jr., "Strawberry Fields Forgotten?" *Chicago Sun-Times*, May 11, 1997, 66.
26. Anonymous industry representative, personal communication, December 2001.
27. For CFCs as well as for all other chemical groups listed a small exception is made for essential uses.

Chapter 5 Unconventional Behavior on Forests

This chapter draws from and expands upon Davenport 2005.

1. See, e.g., Humphreys 1993, 1996; and Porter and Brown 1996.
2. US government, *An International Convention on the World's Forests*, draft of July 5, 1990. This statement was based on FAO assessments of the world's tropical forests from 1980 and 1990 (Sullivan 1993, 159).
3. Report of the Independent Review 1990, 19, cited in Kolk 1996b, 154.
4. United Nations General Assembly, "Resolution Adopted by the General Assembly [on the report of the Second Committee (A/44/746/Add.7)]," A/RES/44/228. 22. March 1990.
5. Panjabi 1997, 144. None of the three assertions in this claim is correct. While the United States was the first state to propose a convention, the idea had already been mooted. Further, the United States' proposal neither focused on tropical timber nor called for any ban on logging.
6. Svensson (1993, 174) gives credit to Sweden for "launch[ing] an initiative on a global forest convention" in 1991, calling it a "Swedish initiative," with no acknowledgment of any of the proposals that were made in 1990.
7. An earlier proposal made by the European Council only called for a protocol to the climate change convention then under negotiation, which would only cover tropical forest protection.
8. In C. Rankin and M. M'Gonigle 1991, "Legislation for Biological Diversity: A Review and Proposal for British Columbia," *U.B.C. Law Review* 277 at 303; quoted in Hughes 1996, 105, fn 236.
9. US official, personal communication, August 2001.
10. Industry spokesperson, personal communication, December 2001.
11. US government official, personal communication, August 2001; US government official, personal communication, August 2001; industry official, personal communication, December 2001.
12. Society of American Foresters 1999, 46.
13. USDA Forest Service 2004, 46.
14. US official, personal communication, August 2001.
15. USDA Forest Service 2004, 43.
16. US Census Bureau 2004–2005, 571. Interestingly, even recent US support for international initiatives on forest law enforcement and governance (FLEG) is qualified by the fact that, according to news reports, American industry lobbyists in the United States have resisted moves to certify that timber is legitimately produced and officials in the State Department worked to defeat restrictions on

timber purchasing being promoted by Britain during its G-8 presidency in 2005. See Paul Brown and Roger Harrabin, "US Tries to Sink Forests Plan: British Initiative on Illegal Logging Opposed," *Guardian*, March 16, 2005, 15.

17. US industry representative, personal communication, December 2001.
18. See, e.g., *Toronto Star*, August 25, 2001, 03; *CanWest News Service*, May 28, 2003, D1.
19. US official, personal communication, August 2001.
20. American Forest and Paper Association (AF&PA) 2002.
21. US official, personal communication, August 2001.
22. United States 1990a.
23. Craig Welch, "A Brief History of the Spotted Owl Controversy," *Seattle Times*, August 6, 2000, A12.
24. Anthony Juniper, of Friends of the Earth (London); quoted in Michael McCarthy, "Counting the Trees Brings Confusion," *Times* (London), June 2, 1992, 10.
25. Porter and Brown 1996, 126.
26. US industry representative, personal communication, December 2001.
27. US official, personal communication, June 2005.
28. US official, personal communication, August 2001; US official, personal communication, June 2005.
29. See David Runnalls, "Bush Rio Talk to Stress Forest," *Earth Summit Times*, June 1, 1992, 1; also see Agarwal, Narain, and Sharma 1999, 227; Johnson 1993, 103; Taib 1997, 78; Kolk 1996b, 155).
30. US official, personal communication, August 2001.
31. Panjabi 1997, 171.
32. US official, personal communication, August 2001.
33. White House Press Release, July 9, 1990.
34. William K. Reilly was director of the EPA and head of the American delegation.
35. US official, personal communication, August 2001.
36. Peter I. Hajnal, ed., *The Seven-power Summit: Documents from the Summits of Industrialized Countries/Supplement—Documents from the 1990 Summit* (Millwood, NY: Kraus International Publications, 1991), 52; cited in Kolk 1996a, 145.
37. US official, personal communication, August 2001.
38. Ibid.
39. The distinction between "forestry" and "forest" or "forest management" in other contexts has become highly politicized in the years since the convention was first proposed. However, in 1990, when forests were only just coming into the international political forum, many in the US government used the terms interchangeably (incorrectly) (US official, personal communication, October 2001).
40. US Talking Points on the Forestry Convention (unpublished).
41. US official, personal communication, October 2001.
42. Fauziah Mohd Taib served on Malaysia's delegation to the UNCED-related negotiations.
43. "Too Much, Too Fast," *Newsweek*, June 1, 1992, 34.

44. John Mukela, "Forests: In Search of Principles," *Developmental Forum*, May–June 1992, 11.
45. Quoted in Sergio Federovisky, "Crece el Optimismo: EEUU Daría Marcha Atrás y Firmaría," *Terra Viva*, June 14, 1992, 11.
46. In Sailesh Kottary, "Indian Minister Criticizes Forest Text," *Earth Summit Times* June 9, 1992, 15; also see Noel L. Gerson, "North-South Agreement on Forests," *Earth Summit Times*, June 13, 1992, 9.
47. US official, personal communication, August 2001. To be fair, it was actually the existence of high standards in the United States, particularly the Endangered Species Act, that caused the controversy to arise, because it prevented the government from being able to lease its old-growth forest lands to industry without public input; this eventually allowed a more equitable stakeholder compromise.
48. US official, personal communication, August 2001.
49. Introductory Statement for G-77 Meeting, August 25, 1990; quoted in Taib 1997, 79.
50. The FAO had also offered earlier to provide the forum for UNCED-related forest negotiations, attempting to legitimize its role as the only international organization mandated to deal with global forest issues. Despite the problems of working within the UNCED process, this offer was rejected, due to the poor reputation the FAO had established with its TFAP program (Humphreys 1996a, 85–87).
51. "Possible Main Elements of an Instrument (Convention, Agreement, Protocol, Charter, etc.) for the Conservation and Development of the World's Forests" (FAO draft), Rome, September 18, 1990.
52. PrepCom 2 also took a decision that firmly placed the remainder of the forest negotiations within the UNCED process and precluded any possible shift to the FAO.
53. The concept of "opportunity cost foregone" appeared many times during the UNCED forest debates, beginning with FAO documents prior to PrepCom 2 (e.g., COFO-90/3(a), paragraph 34); see Humphreys 1996a, 91, fn 33.
54. US Statement on Forests to PrepCom 2, Geneva, March 25, 1991.
55. My italics.
56. See "PrepCom II: What Was Accomplished?" *Earth Summit Update No. 1.*, July 1991, 1.
57. This view has been expressed since then as well, and in other fora, such as in the ITTO (personal observation; delegate to the ITTO, personal communication, May 1995).
58. UN document A/CONF.151/PC/WG.I/Misc.3, "*Ad Hoc* Subgroup, Forests, Draft Synoptic List: Compiled by the Secretariat on the Basis of Informal Consultations," March 26, 1991, 4; cited in Humphreys 1996a, 91.
59. UN document A/CONF.151/PC/WG1/L.18 Rev.1, "Revised Decision Submitted by the Chairman on the Basis of Informal Consultations," paragraph 5.3; quoted in Humphreys 1996a, 93.
60. "G-7, Developing Country Communiqués Highlight UNCED Conflicts," *Earth Summit Update No. 2.*, August 1991, 2.

61. "US Principles Downplay Primary Forests," *Earth Summit Update No. 1.*, July 1991, 1.
62. US proposal, "Principles for a Global Forest Convention/Agreement"; quoted in Humphreys 1996a, 94.
63. "US Will Oppose Key Agenda 21 Options at PrepCom 3," *Earth Summit Update No. 2.*, August 1991, 1, 2.
64. UN document A/CONF.151/PC/65, "Guiding Principles for a Consensus on Forests"; cited in Humphreys 1996a, 93.
65. UN document A/CONF.151/PC/WG.I/L.22, Proposal Submitted by Ghana (on Behalf of the Group of 77), August 16, 1991; cited in Humphreys 1996a, 94.
66. A/CONF.151/PC/WG.I/L.22, principles 12 and 13, 3; cited in Humphreys 1996a, 97.
67. This was defined as "a commitment to provide new and additional financial resources to developing countries, for meeting, *inter alia*, the commitments under Agenda 21, and other sustainable development concerns" (UN document A/CONF.151/PrepCom/L.41, "China and Ghana (on behalf of the Group of 77): Draft Decision: Financial Resources," 2; quoted in Humphreys 1996a, 97.
68. UN document A/CONF.151/PC/86, "Proposal submitted by the Delegation of the People's Republic of China, The Green Fund." August 15, 1991; cited in Humphreys 1996a, 96. It was envisaged that this Fund would comprise equitable representation from developed and developing countries, deal with local problems in developing countries such as forest preservation and tree planting, be funded primarily by developed countries and international agencies, and provide disbursement to all developing countries without conditionality.
69. Term used by UNCTAD official, personal communication, January 1994.
70. Nicholas van Praag, World Bank; quoted in Joy Elliott, "Jury Undecided on GEF's First Year," *Earth Summit Times*, June 2, 1992, 3.
71. Ibid.
72. US official, personal communication, August 2001.
73. UN document A/CONF.151/PrepCom/WG1/CRP.14/Rev.2, "A Non-Legally Binding Authoritative Statement of Principles for a Global Consensus on the Management, Conservation and Sustainable Development of all Types of Forests," September 3, 1991 (disputed language is in brackets); quoted in Taib 1997, 82.
74. "US Rejects Targets in Forest Negotiations," *Earth Summit Update*, September 1991, 2.
75. Quoted in Humphreys 1993, 50.
76. Ibid.
77. Negotiations on the Agenda 21 chapter on forests did not actually begin until PrepCom 4 ("Issue-by-Issue Summary of PrepCom 3," *Earth Summit Update*, Special Supplement, October 1991).
78. US official, personal communication, August 2001.
79. Statement by Ambassador Ting Wen-Lian at UN Briefing for the Press, Rio de Janeiro, June 2, 1992; quoted in Taib 1997, 83.
80. US official, personal communication, August 2001.

81. Dr. Mahathir Bin Mohamad, Malaysian prime minister; quoted in Humphreys 1996a, 101.
82. Grubb et al. 1993, 36, fn 15.
83. Kuala Lumpur Declaration on Environment and Development, April 29, 1992.
84. US official, personal communication, August 2001.
85. "US Rejects Targets in Forest Negotiations."
86. US chief negotiator Curtis Bohlen, in John Vidal and Paul Brown, "Deadlock in Talks about Aid Cash," *Guardian*, June 10, 1992, 8.
87. James Gerstenzang, "Bush Proposes Huge Growth In Forest Funding," *Los Angeles Times*. June 2, 1992, A10.
88. "Un Susurro en el Bosque," *Crosscurrents*, June 5, 1992, 22; see also Sergio Federovisky, "EEUU Quiere Reabrir el Debate," *Terra Viva*, June 3, 1992, 2; Earl Lane, "Proposal Doubles Money for Forests," *Newsday*, June 2, 1992, 79.
89. After a pilot program in which around $9 million was allocated to three countries, the Forests for the Future initiative dried up under the Clinton administration (US official, personal communication, August 2001).
90. One US official who was involved in the calculations that went into the Forests for the Future proposal has assured me that it did represent new appropriations from Congress to USAID, the US Forest Service, and the EPA for global forest project funding (personal communication, October 2001).
91. Porter and Brown 1996, 199, fn 82.
92. US official, personal communication, August 2001.
93. See Taib 1997, 85; see also "US Singled Out as Eco Bad Guy," *Jornal do Brasil*, special edition, June 5, 1992, 1, 6.
94. See Joy Elliott, "Nations Agree to Draft Desert Convention," *Earth Summit Times*, June 11, 1992, 3, for a similar report.
95. David E. Pitt, "US Pulls All Stops for 'Forest Principles'," *Earth Summit Times*, June 10, 1992, 1, 16.
96. Thalif Deen, "Desertificacion y bosques: EEUU Esperaba un Canje?" *Terra Viva*, June 12, 1992, 9.
97. Chakravarthi Raghavan, "Stumbles on Finance, Forests, Air and Deserts," *SUNS (South North Development Monitor) at the Earth Summit, No. 3.*, June 10, 1992, 1–2.
98. Discussion on another contentious issue—whether a developed country commitment to devote 0.7% of their GNP to official development assistance (ODA), made over 20 years earlier, should have an explicit target deadline—did not include the United States as it had never agreed to that commitment originally, its ODA figure being the lowest of the developed countries at 0.15%. See "US May Veto Key Parts of Agenda 21," *Earth Summit Update No. 7.*, March 1992, 1, 3; Thalif Deen, "Mecanismos de Financiamiento Dividieron al Grupo de los 77," *Terra Viva*, June 9, 1992, 4.
99. David E. Pitt, "Forest . . . Finance . . . Frustration," *Earth Summit Times*, June 12, 1992, 1, 16.
100. "Nations Have Sovereignty over Forests," *Jornal do Brasil*, special edition, June 13, 1992, 6.

101. Of course, this also meant that the South lost its leverage over Agenda 21, with the result that Agenda 21 did not reflect the South's positions, on finance in particular.
102. UN document A/CONF.151/6/Rev. 1 (the Forest Principles, final version), Preamble (d), 1.
103. UN document A/CONF.151/6/Rev. 1, quoted in Kolk 1996a, 14.
104. Heissenbuttel et al. 1992, 16.
105. US official, personal communication, October 2001.
106. US industry representative, personal communication, December 2001; also see Heissenbuttel et al. 1992.
107. Total costs of the financial requirements for environmental protection in developing countries were estimated at $600 billion, of which $125 billion would have to come from donor countries—a figure that happened to be equivalent to 0.7% of the GNP of the OECD nations (Kolk 1996a, 16).
108. David Runnalls, "Summit Recap: No Cash, More G-77," *Earth Summit Times*, June 14, 1992, 3.
109. The United States was on the opposing side of this dispute within the Agenda 21 negotiations; see Daniel R. Abbasi, "Agenda 21 Disputes Are on the Table," *Earth Summit Times*, June 11, 1992, 3.
110. Michael Howard, British Secretary of State for the Environment, June 24, 1992; quoted in Johnson 1993. It should be noted that Sullivan calls them a "modest step forward" (167).
111. US official, personal communication, December 2001.
112. US NGO participant, personal communication, August 2001.
113. "Nations Have Sovereignty Over Forests, 1992."
114. US official, personal communication, August 2001.
115. Harkavy 1992, 12.
116. Grubb et al. 1993, 36, fn 15. Taib also makes this point (1997, 58).
117. Lane 1992, 79.
118. Ambassador Bernardo Pericas, head Brazilian forest negotiator, quoted in Maria Elena Hurtado, "Northern Obsession with Forests Boggles the Mind," *Crosscurrents*, June 5, 1992, 12.
119. Kolk (1996a) sees the softening of the Malaysian position as the result of the retirement of Madam Ting, its "hard-line representative" in Rio, and the interministerial shuffle of Malaysia's forest portfolio from the Ministry of Foreign Affairs to Primary Resources, which depoliticized the issue. Malaysia claimed that its shift was due to its new efforts toward producing "environmentally correct" wood (Agarwal, Narain, and Sharma 1999, 244). The primary factor, however, was probably the cooperation that grew between the governments of Canada and Malaysia after UNCED that put to rest Malaysia's original fear that a convention would legitimize trade discrimination with a new conceptualization of a convention as legitimizing the trade—perhaps at a low standard of "sustainability" (US official, personal communication, June 1997).
120. Porter and Brown 1996, 118.
121. Senator Albert Gore and Congressman John Porter, Joint Resolution Calling on the President of the United States to Take a Leadership Role in the International

Negotiations Toward a World Forest Convention and a Framework Convention on Climate Change, and for Other Purposes (S.J. Res. 181/H.J. Res. 302), Washington, DC.

122. Mukela 1992.
123. Angela Harkavy, "CAPE '92" (US NGO network), quoted in David E. Pitt, "US Pulls All Stops for 'Forest Principles'."
124. In the end, the United States promised $250 million in new funding for all of the actions identified at UNCED (David Runnalls, "Summit Recap: No Cash, More G-77").
125. President Bush himself called for a contribution of $24 billion in Western aid to Russia just before Rio (Richard Nixon, "Yeltsin Clearly Deserves Help," *Earth Summit Times*, June 13, 1992, 15).
126. Fearon 1991, 183, fn 35.
127. As Fearon puts it, the generic counterfactual proposition is "If it had not been the case that C (or not C), it would have been the case that E (or not E)" (1991, 169).
128. Even in the case of the MLF, which the United States did not initiate, US agreement to it was necessary before the anti coalition was willing to come into the Montreal Protocol.
129. In other words, the value of nonagreement to Malaysia was a bit higher than it would have been otherwise, and therefore an agreement would have to be more valuable to overcome the value of the no agreement outcome.
130. Indeed, one US official recalls that it was the United States' desire for a convention that may have been the most responsible for US opposition to expanding the scope of the ITTA when it was being renegotiated in 1993–1994 (US official, personal communication, October 2001).
131. US official, personal communication, August 2001.
132. US official, January 1994; US official, August 2001; Agarwal, Narain and Sharma 1999, 244.
133. US official, personal communication, August 2001.
134. US official, personal communication, August 2001.
135. The concern that it is the major timber exporting countries have pushed for the convention since Rio has been echoed by many others, including US government officials (personal communication, June 1997).
136. US NGO representative, personal communication, August 2001.
137. US official, personal communication, September 1996.
138. In this context, "open-ended" means that any country may participate.
139. Quoted in Carpenter et al. 1997, 6.
140. The relationship of the CBD to the IFF, for instance, was highly contentious. For example, in 1997 Jaime Hurtubia of the IFF Secretariat made a presentation that infuriated some NGOs as it seemed to suggest IFF supremacy over the CBD on forest biology even though the IFF was only an *ad hoc* UN body set up under the authority of no binding agreement (Presentation by Mr. Jaime Hurtubia, IFF Secretariat, to the Third Meeting of the Subsidiary Body on Scientific, Technical and Technological Advice of the Convention on Biological

Diversity, September 1–5, 1997, Montreal, Canada, quoted in William E. Mankin (1998), "Entering the Fray; International Forest Policy Processes: An NGO Perspective on their Effectiveness," Discussion paper). London: International Institute for Environment and Development.

141. After 1997 the United Kingdom in particular distanced itself from the call for a convention (personal communication, U.K. official, June 1997; see also Humphreys 1996b, 162; Dimitrov et al. 2000, 11, on the EU's view in 2000).
142. United Nations (2000), "Advance Unedited Text of the Report of the Intergovernmental Forum on Forests at its Fourth Session" (IFF Secretariat, New York: IFF Secretariat); quoted in Dimitrov et al. 2000, 12.
143. UN 2000 (ibid.); quoted in Humphreys 2001, 164.
144. US official, personal communication, December 1999.
145. This is not surprising; another US official comments that "the US generally gets its way" (personal communication, August 2001). Again, though, what this means depends on the costs and benefits of what the United States wants, as demonstrated by the Rio outcome on forests.
146. US official, personal communication, August 2001.
147. Ibid.
148. US official, personal communication, August 2001; US official, personal communication, August 2001; NGO representative, August 2001; Agarwal, Narain, and Sharma 1999.
149. US official, personal communication, August 2001.
150. US industry representative, personal communication, December 2001.
151. Observer at UNFF-5, personal communication, May 2005.
152. Industry spokesperson, personal communication, April 2005.
153. Delegates to UNFF-4 and UNFF-5, personal communications, May 2004, May 2005.
154. Encouragement of substitutes has never been raised as an objective for a convention or in the Forest Principles, although substitutes exist for some uses, such as steel framing for housing.
155. See chapter 3 on the original Toronto Group and EC proposals on ozone. Another example of this appeared in the consumer countries' side statement negotiated alongside the ITTA 1994, in which tropical-timber consuming countries pledged to do exactly what they were already committed to doing (personal observation; US official, personal communication, December 2001; British official, personal communication, April 2006).

Chapter 6 The Climate on Climate Change

1. Andrew Cave, "Katrina's Trail of Destruction Spiralling Towards $100bn," *Daily Telegraph*, September 5, 2005, 27.
2. "Unnatural Disaster—The Lessons of Katrina," Worldwatch Institute Press Release. September 2, 2005. Web address: http://www.worldwatch.org/ct/20050902/press/news/2005/09/02/.

3. Eric Berger, "Keeping its Head Above Water: New Orleans Faces Doomsday Scenario," *Houston Chronicle*, December 1, 2001, A29.
4. Dominic Izzo, "Reengineering the Mississippi," *Civil Engineering Magazine*, July 2004. Web address: http://www.pubs.asce.org/ceonline/ceonline04/0704feat.html.
5. Recent research shows the intensity of hurricanes, including wind speed and duration, seems to have risen by about 70% in the past 30 years. See P. J. Webster, G.J. Holland, J.A. Curry, and H.-R. Chang, "Changes in Tropical Cyclone Number, Duration, and Intensity in a Warming Environment," *Science* 309 (September 15, 2005): 1844–1846; Alistair Doyle, "Did Climate Change Drive Katrina?" Australian Broadcasting Company, September 12, 2005. Web address: http://www.abc.net.au/science/news/stories/s1458489.htm.
6. For this reason, the original term "global warming" has been superseded by the term "climate change" to label the full phenomenon.
7. UN Framework Convention on Climate Change: The Greenhouse Effect and the Carbon Cycle. Web address: http://unfccc.int/essential_background/feeling_the_heat/items/2903.php.
8. In this I follow Bodansky (2001), among others. Bodansky, however, labels this the "global climate *change* regime." Given that the norms, principles, rules, and decision-making procedures of this regime are intended to protect the climate, not the phenomenon of climate change, I prefer "global climate regime."
9. This list does not include GHGs that are already covered by the ozone agreements, such as CFCs.
10. Unless otherwise noted, all article numbers in this section refer to the Kyoto Protocol.
11. Lomborg 2001, 259.
12. The Intergovernmental Panel on Climate change called for a 60% cut in its First Assessment Report in 1990.
13. The target for each of these countries in Annex B of the Kyoto Protocol, "Party Quantified Emission Limitation" is 108%, 110%, and 101% of 1990 emissions levels, respectively (Kyoto Protocol, Annex B).
14. United Nations Framework Convention on Climate Change. Decision 27/CMP.1: Procedures and Mechanisms Relating to Compliance under the Kyoto Protocol. In the Report of the Conference of the parties Serving as the Meeting of the Parties to the Kyoto Protocol on its First Session, held at Montreal from November 28 to December 10, 2005, Addendum: Action Taken by the Conference of the Parties Serving as the Meeting of the Parties to the Kyoto Protocol at its First Session, 2005b (FCCC/KP/CMP/2005/8/Add.3). Web address: http://unfccc.int/resource/docs/2005/cmp1/eng/08a03.pdf#page=92; United Nations Framework Convention on Climate Change, Kyoto Protocol to the United Nations Framework Convention on Climate Change, 1997 (FCCC/CP/1997/L.7/Add.1). Web address: http://unfccc.int/cop3/107a01. htm.
15. Richard Black, "Will Kyoto Die at Canadian Hands?" BBC News, January 27, 2006. Web address: http://news.bbc.co.uk/go/pr/fr/-/2/hi/science/nature/4650878.stm.

16. UNFCCC website. Web address: http://unfccc.int/files/essential_background/convention/status_of_ratification/application/pdf/ratlist.pdf.
17. UNFCCC website. Web address: http://unfccc.int/files/essential_background/kyoto_protocol/application/pdf/kpstats.pdf.
18. Figures taken from Climate Change: PRC Officials Comment At US, PRC, Japan Tech Transfer and GHG Mitigation NGO Seminar: A Report from Embassy, Beijing, December 1997. Web address: http://www.usembassy-china.org.cn/sandt/clmoffic.htm.
19. Monastersky 1990b, 263. Eleven years later, George W. Bush felt confident enough to ignore this warning and question climate change science (see Nancy Dunne, "Bush backs Away from Pledge to Curb Carbon Emissions," *Financial Times*, March 14, 2001, 1).
20. Michael Weisskopf, "US Gets Mixed Reviews On Global Warming Plan; 'Action Agenda' Lacks Carbon Dioxide Target," *Washington Post*, February 5, 1991, A3.
21. "Climate Convention Negotiations Face Difficulties, But 1992 Target Feasible, Chairman Says," *United Nations Chronicle*, 02517329, June 1991, 56–57.
22. Quoted in Fred Pearce, "Forum: Captain Eco Rides Again," *New Scientist*, October 4, 1997, 4646.
23. Michael Prowse and David Lascelles, "Financing a Green Future in a Planet without Borders," *Financial Times*, February 14, 1992, 5.
24. Alliance of Small Island States website. Web address: http://www.sidsnet.org/aosis/.
25. Paul Brown, "Way Open for Cuts in Greenhouse Gases; Breakthrough Clears a Path for Hard Bargaining on Targets and Implementation Timetables," *Guardian*, April 8, 1995, 12.
26. Senate Debate over the Byrd-Hagel Resolution, July 25, 1997. In David G. Victor. *Climate Change: Debating America's Policy Options*, 2004 (New York: Council on Foreign Relations Press, 2004), Appendix A, 117–129. Council on Foreign Relations.
27. Senator John Kerry, interview with author, Kyoto, Japan, December 9, 1997. Available at http://www.iisd.ca/climate/kyoto/coverage.html.
28. Senator Trent Lott, quoted in Expressing Sense of Senate Regarding UN Framework Convention on Climate Change (Senate—July 25, 1997). Congressional Record: July 25, 1997 (Senate), S8113–S8139. From the Congressional Record Online via GPO Access, DOCID:cr25jy97–97. Web address: wais.access.gpo.gov.
29. Ross Gelbspan, "Bush's Gambit on Climate," *Christian Science Monitor*, April 2, 2001, 9.
30. Senator John Kerry, interview with author, Kyoto, Japan, December 9, 1997. Available at http://www.iisd.ca/climate/kyoto/coverage.html.
31. As of 2005, the idea that the time for finger-pointing was over was still not universally accepted, nor were North-South equity concerns being voiced solely by developing countries. A group of nine institutions from around the world put together a presentation for a side event at UNFCCC COP-11/MOP-1 in

December 2005, to consider questions such as responsibility and liability for damages from climate change, atmospheric targets, and allocation of GHG emissions reductions (Rock Ethics Institute, 2005).

32. Senator Chuck Hagel, quoted in Expressing Sense of Senate Regarding UN Framework Convention on Climate Change (Senate—July 25, 1997). [Congressional Record: July 25, 1997 (Senate), S8115. From the Congressional Record Online via GPO Access. DOCID:cr25jy97–97. Web address: wais.access.gpo.gov.
33. Senator Robert Byrd, quoted in Expressing Sense of Senate Regarding UN Framework Convention on Climate Change (Senate—July 25, 1997). Congressional Record: July 25, 1997 (Senate), S8117. From the Congressional Record Online via GPO Access. DOCID:cr25jy97–97. Web address: wais.access.gpo.gov.
34. Federal Reserve Chairman Alan Greenspan, for instance, noted in 2004 that "there is palpable unease that businesses and jobs are being drained from the United States;" quoted in "David Ignatius, Dishonest Trade Talk," *Washington Post*, February 24, 2004, A21.
35. See, e.g., "Kyoto is Dead Duck until US Clambers on Board," *Insurance Day*. February 16, 2005. Web address: http://www.insurancedat/insday/homepage. jsp?pageid=article&articleid=20000068436.
36. All of these alternatives are listed in Matthew Paterson, "Global Warming." In Caroline Thomas, ed., *The Environment in International Relations* (London: The Royal Institute of International Affairs, 1992) chapter 5, 155–198; 88–89.
37. Personal observation, UNFCCC COP-3, Kyoto, Japan, December 11, 1997; Bettelli et al., 1997, 15. Kyoto Protocol to the United Nations Framework Convention on Climate Change (FCCC/CP/1997/L.7/Add.1) and Kyoto Protocol to the United Nations Framework Convention on Climate Change: Final Draft by the Chairman of the Committee of the Whole (FCCC/CP/1997/CRP.6). The text of the draft article allowed any non-Annex I party to undertake formally a level of limitation or reduction of anthropogenic GHG emissions of its choosing, based on a base year or period of its choosing, and to opt to be bound by such a formal declaration.
38. Senator Robert Byrd. In Senate Debate: Global Climate Change: The Kyoto Protocol. Congressional Record: January 29, 1998, S197. From the Congressional Record Online via GPO Access, DOCID:cr29ja98-20. Web address: wais.access.gpo.gov.
39. Ibid.
40. NGO observers at UNFCCC COP-3, personal communications, December 1997; personal observation.
41. Byrd, Global Climate Change.
42. Senator Larry E. Craig. In Senate Debate: U.N. Global Climate Treaty, Congressional Record: April 20, 1998, S3244. Congressional Record Online via GPO Access, DOCID:cr20ap98–17. Web address: http://frwebgate.access.gpo.gov/ cgi-bin/ getpage.cgi? position = all&page= s3244&dbname= 1998_ record. This was still somewhat lower than the $800 billion that the President's Council

of Economic Advisors had predicted in 1990 that it would cost, at a minimum, to cut the United States' carbon emissions by 20% by the year 2100. US Council of Economic Advisors, *Economic Report of the President* (Washington, DC: US Government Printing Office, 1990), 214; cited in Rowlands 1995, 134.

43. Eric Pianin, "US Aims to Pull Out of Warming Treaty; 'No Interest' in Implementing Kyoto Pact, Whitman Says," *Washington Post*, March 28, 2001, A01.
44. John R. Justus and Susan R. Fletcher. CRS Issue Brief for Congress. IB89005: Global Climate Change. Web address: http://www.ncseonline.org/NLE/CRSreports/Climate/clim-2.cfm?&CFID=1732361&CFTOKEN=90447656#_1_12.
45. William D. Nordhaus, "Economic Approaches to Greenhouse Warming." In Rudiger Dornbusch and James M. Poterba, eds., *Global Warming: Economic Policy Responses* (London: The MIT press, 1991); quoted in Rowlands 1995, 138.
46. William R. Cline, *The Economics of Global Warming* (Washington, DC: Institute for International Economics, 1992); quoted in Rowlands 1995, 141.
47. Brazilian delegate, UNFCCC COP-3, personal communication, December 1997.
48. National Academy of Sciences, *Policy Implications of Greenhouse Warming* (Washington, DC: NAS Press, 1991), 106; quoted in Rowlands 1995, 139.
49. Michael Oppenheimer and Robert Boyle, *Dead Heat: The Race Against the Greenhouse Effect* (London: I.B. Tauris, 1990), 164; quoted in Rowlands 1995, 139.
50. Lawrence Summers of the World Bank, e.g., called for a discount rate of at least 8% in studies regarding climate change (Rowlands 1995, 140).
51. William D. Nordhaus and Joseph G. Boyer, "Requiem for Kyoto: An Economic Analysis," *Energy Journal Special Issue* (1999): 93–130; and Richard S.J. Tol, "Kyoto, Efficiency, and Cost-Effectiveness: Applications of FUND," *Energy Journal Special Issue* (1999):131–156; cited in Warwick J. McKibbin and Peter J. Wilcoxen, "The Role of Economics in Climate Change Policy," *Journal of Economic Perspectives* 16 (2002):107–130.

Chapter 7 Conclusion: A Climate for Future Success?

1. It is also interesting to note that the climate change case also has the issue of sovereignty over natural resources in common with the forest case, illustrated most forcefully by China's unwillingness to give up its sovereignty in making decisions on use of its vast resources of "dirty coal" that produces huge amounts of CO_2 and other pollutants. See Mark Clayton, "New coal plants bury 'Kyoto,' " *Christian Science Monitor*, December 23, 2004, 1; Margaret Kriz, "Fueling the Dragon," *National Journal*, August 6, 2005, 2510–2513.
2. Some scholars may take issue with the description of the climate change and deforestation cases as "failed" because both are the subject of an international cooperative regime. I use the term to emphasize that a low level of cooperation is insufficient if it cannot prevent a harmful outcome from continual environmental degradation. By my definition of effectiveness as well as with regard to the deeper question of

actual improvement in the environmental conditions that originally prompted the cooperative efforts, international arrangements to address these cases have so far failed.

3. See, e.g., materials on the Costa Rica-Canada Initiative (Final Meeting December 6–10, 1999). Web address: http://www.iisd.ca/sd/crci/final/.
4. Valerie Lawton, "Canada Stays Calm On Kyoto Pullout," *Toronto Star*, March 30, 2001.
5. See Scott Barrett, "Kyoto's Fall". *German-American Relations and the Presidency of George W. Bush*, American Institute for Contemporary German Studies, 2002. Web address: http://www.aicgs.org/aicgs/aicgs/aicgs/aicgs/aicgs/aicgs/aicgs/aicgs/aicgs/research/g2001/barrett.shtml.
6. Indeed, in his documentary, *An Inconvenient Truth*, released in 2006, Gore admits that even had the Senate been controlled by Democrats in 1998, the result, in terms of the Senate's rejection of the Kyoto Protocol, "would have been nearly the same." Quoted in Bret Schulte, "Saying It in Cinema," *US News & World Report*, June 5, 2006, 39.
7. United Nations Environment Programme, Global Environment Outlook, "Climate Change and Extreme Events," in *GEO Year Book 2006*. Web address: http://www.unep.org/geo/yearbook/yb2006/009.asp.
8. Steve Connor, "The Final Proof: Global Warming is a Man-Made Disaster," *Independent*, February 19, 2005, 1.
9. Geoffrey Lean, "Global Warming Will Plunge Britain into New Ice Age 'Within Decades'," *Independent*, January 25, 2004, 18.
10. "The Most Important Issue that We Face," *Independent*, April 18, 2005, 1.
11. Michael McCarthy, "Climate Change Poses Threat to Food Supply, Scientists Say," *Independent*, April 27, 2005, 11.
12. Andrew C. Revkin, "In a Melting Trend, Less Arctic Ice to Go Around," *New York Times*, September 29, 2005, 1; Steve Connor, "Global Warming 'Past the Point of No Return'," *Independent*, September 16, 2005, 1–2.
13. Charles Clover, "Why the Inuit People are Walking on Thin Ice," *Daily Telegraph*, October 17, 2005, 13.
14. Andrew Buncombe, "Global Warming: Will You Listen Now, America?" *Independent*, August 19, 2005, 1–3.
15. Kate Ravilious, "Global Warming: Death in the Deep-Freeze," *Independent*, September 28, 2005, 44–45.
16. Danial R. Abbasi, *Americans and Climate Change, Closing the Gap between Science and Action: Synthesis of Insights and Recommendations from the 2005 Yale Forestry and Environmental Studies Conference on Climate Change* (New Haven, CT: Yale School of Forestry and Environmental Studies, 2006).
17. Laurie Goodstein, "Evangelical Leaders Join Global Warming Initiative," *New York Times*, February 8, 2006, A12. This trend toward increasing recognition of environmental costs may be bolstered by continually increasing evidence of global warming and its effects, such as reports that Greenland ice cap melting could now be approaching a tipping point toward "explosively rapid" disintegration,

which appeared almost simultaneously with the announcement of the Evangelical Climate Initiative; see Jim Hansen, "Greenland Ice Cap Melting at Twice the Rate It Was Five Years Ago, Says Scientist Bush Tried to Gag," *Independent*, February 17, 2006, 1.

18. "Climate Change Poses Major Threat to US Insurance Industry," *Insurance Journal*, September 19, 2005. Web address: http://www.insurancejournal.com/magazines/east/2005/09/19/features/60454.htm.
19. "Insurers Link Global Warming with Higher Cost of Storms," *Environment News Service*, July 28, 2005.
20. Diane Levick, "Insurers Tackle Climate; Industry Urged to Seek New Types Of Coverage," *Hartford Courant* (Connecticut), October 28, 2005, E1.
21. Dean Starkman, "A New Worry for Insurers; Firms Looking at Whether Climate Change Could Affect Their Bottom Lines," *Washington Post*, October 5, 2005, D1.
22. Climate Change Poses Major Threat to US Insurance Industry," *Insurance Journal*, September 19, 2005.
23. "Senior UN Officials, Pension Fund Heads, CEOs, Wall Street Leaders to Discuss Climate Risks, Opportunities at Summit, 10 May," UN Press Release Note No. 5938, May 6, 2005.
24. "Thirteen Pension Leaders Call on SEC Chairman to Require Global Warming Risks in Corporate Disclosures," Ceres Press Release, April 15, 2004.
25. UN Press Release Note No. 5938.
26. UNCTAD representative, personal communication, May 26, 2005.
27. Jeffrey Immelt, "Ecomagination," Statement made May 9, 2005; "GE Launches Ecomagination to Develop Environmental Technologies; Company-Wide Focus on Addressing Pressing Challenges." Press Release, General Electric Company, May 9, 2005.
28. "Climate Signals," *New York Times*, May 19, 2005, 26.
29. GE representative, personal communication, August 17, 2005.
30. "Climate Signals."
31. Vanessa Houlder, "Climate Change Could Be Next Legal Battlefield," *Financial Times*, July 14, 2003, 10.
32. See "Climate Justice: Enforcing Climate Change Law," 2005. Web address: http://www.climatelaw.org/.
33. Paul Brown, "Rich Nations Could be Sued by Climate Victims," *Guardian*, July 10, 2001, 7.
34. David Adam, "50m Environmental Refugees by End of Decade, UN Warns: States Urged to Prepare for Victims of Climate Change: Natural Disasters Displace More People than Wars," *Guardian*, October 12, 2005, 24.
35. "Nuisance case: States and NGOs Appeal Dismissal" (December 15, 2005). Climate Justice Programme, 2005/12/22. Web address: http://www.climatelaw.org/media/US%20nuisance%20appeal.
36. Catherine Trevison, "Suit over Emission Threat Hovers in Legal Gray Zone," *Oregonian*, October 5, 2005, C01.

37. Jane Kay, "State Seeks 30% Cut in Tailpipe Emissions; Automakers Say They May Sue If 10-Year Plan OKd," *San Francisco Chronicle*, June 15, 2004, A1.
38. David Gram, "Vermont Adopts New Rules to Cut Car CO_2 Emissions: New Standards Take Effect for 2009 Model Year," *Detroit News Autos Insider*, November 3, 2005. Web address: http://www.detnews.com/2005/autosinsider/0511/03/0auto-370678.htm.
39. Polly Ghazi, "A Storm Brewing: Is Bush Out of Step with US Public Opinion?" *Guardian*, July 6, 2005, 13, John Holusha, "California Bill Calls for Cuts in Emissions," *New York Times*, April 4, 2006, A19.
40. "US States Say Power Bills Won't Soar on CO_2 Plan," Reuters, November 7, 2005. Web address: http://today.reuters.com/investing/financeArticle.aspx?type=bondsNews&storyID=URI:urn:newsml:reuters.com:20051107:MTFH82717_2005-11-07-23-16-47_N07555531:1; Mark Clayton, "One Region's Bid to Slow Global Warming," *Christian Science Monitor*, December 22, 2005, 2. Participating states include Connecticut, Delaware, Maine, New Hampshire, New Jersey, New York, and Vermont. In addition, legislation passed in April 2006 requires Maryland to become a full participant by 2007, and the District of Columbia, Massachusetts, Pennsylvania, Rhode Island, the Eastern Canadian provinces, and New Brunswick are currently observers in the process. See Regional Greenhouse Gas Initiative: An Initiative of the Northeast and Mid-Atlantic States. "Participating States," Web address: http://rggi.org/states.htm.
41. Paul Brown, "US Cities Snub Bush and Sign Up to Kyoto: Dozens of Mayors, Representing More Than 29 Million Americans, Pledge to Cut Greenhouse Gases," *Guardian*, May 17, 2005, 19; Paul H. B. Shin, "White House's Icy Reception to Climate Threats not Stopping Some Cities," *New York Daily News*, December 11, 2005, 26.) This could potentially translate into different standards for firms that do business in different cities.
42. Fred Pearce, "European Trading in Carbon-Emission Permits Begins," *New Scientist*, January 6, 2005; Loren Cass, "Norm Entrapment and Preference Change: The Evolution of the European Union Position on International Emissions Trading," *Global Environmental Politics* 5:38 (2005): 60; Guri Bang, Anreas Tjernshaugen, and Steinar Andresen, "Future US Climate Policy: International Re-engagement?" *International Studies Perspectives* 6 (2005): 285–303; Robert Collier, "State Bypasses Kyoto, Fights Global Warming: California Tries to Cut Emissions on its Own," *San Francisco Chronicle*, February 1, 2005, Al; Jad Mouawad, "No US Rules, But Some Firms Reduce Emissions," *New York Times*, May 29, 2006. Web address: http://www.int.com/articles/2006/05/29/business/carbon.php.
43. Margaret Kriz, "Fueling the Dragon," *National Journal*, August 6, 2005, 2510–2513.
44. Justin Blum, "Making an Oil Pledge; Declaration of Undependence Rests on New Energy Sources," *Washington Post*, February 2, 2006, D1.
45. Immelt, "Ecoimagination"; "Climate Signals".
46. Statement of Cecil E. Roberts, President of the United Mine Workers of America before the United States Senate Committee on Energy and Natural resources on

the Economic Impacts of the Kyoto Protocol, March 25, 1999. Web address: http://www.umwa.org/legaction/globalwarming/statement.shtml.

47. Subsidies for development of alternative energies extend back to the 1970s; the Energy Policy Act of 2005 would provide more than $11 billion in incentives for the development of nonfossil fuel energy, conservation, energy efficiency, and cleaner technologies, mainly in the form of tax breaks. It should be noted, however, that the US government also subsidizes not-so-climate-friendly activities, such as road construction, oil extraction and production, fossil fuel use, and highway patrols See Barnaby J. Feder, "A Different Era for the Alternative Energy Business," *New York Times*, May 29, 2004, C1; Congressional Budget Office, Cost Estimate for the Bill Conference Agreement, H.R. 6, Energy Policy Act of 2005, July 27, 2005. Web address: http://www.cbo.gov/showdoc.cfm?index= 6581&sequence=0&from=6; also see US Congress, Energy Policy Act of 2005 (H.R.6/Public Law 109–58), August 8, 2005. Web address: http:// frwebgate.access.gpo.gov/cgi-bin/getdoc.cgi? dbname=109_cong_bills&docid=f:h6enr.txt.pdf. Also see UNEP 1999a.
48. Statement by ARCO Chief Executive Officer Michael Bowlin, February 1999; quoted in Lester R. Brown, "The Rise and Fall of the Global Climate Coalition," Earth Policy Institute, July 25, 2000. Web address: http://www.earth-policy. org/Alerts/Alert6.htm.
49. Kriz, "Fueling the Dragon"; Lindsey Beck, "China Unlikely to Sign On To Kyoto Emissions Cuts," *Macon (Georgia) Daily Online*, November 12, 2005. Web address: http://www.maconareaonline.com/news.asp?id=12544.
50. See Margaret Kriz, "Heating Up: Global Warming Moves to a Front Burner, as Demands Grow for Aggressive Action to Limit Greenhouse-Gas Emissions," *National Journal*, August 6, 2005, 2504–2508; Ghazi, "A Storm Brewing"; Bang and Froyn 2005; also see Bang, Tjernshaugen, and Anderson 2004; Barkin and DeSombre 2004.
51. Katharine Q. Seelye, "President Distances Himself from Global Warming Report," *New York Times*, June 5, 2002, 23; Antony Barnett, "Bush Attacks Environment 'Scare Stories': Secret Email Gives Advice on Denying Climate Change," *Observer*, April 4, 2004, 22.
52. "Climate Change, Clean Energy and Sustainable Development," Statement of the G-8, Gleneagles 2005. Web address: http://www.number-10.gov.uk/output/Page7881.asp.
53. Gary Younge, "US Official Accused of Doctoring Papers Quits," *Irish Times*, June 13, 2005, 11.
54. See Tarek Maassarani, "Muzzling Scientists on Warming," *Topeka Capital-Journal*, May 19, 2006. Web address. http://www.cjonline.com/stories/051906/opi_maassarani.shtml; Andrew C. Revkin, "Climate Expert Says NASA Tried to Silence Him," *New York Times*, January 29, 2006, 1.
55. Intergovernmental Panel on Climate Change. Report of the Joint IPCC Working Group II and III Expert Meeting on the Integration of Adaptation, Mitigation and Sustainable Development into the 4th IPCC Assessment Report, St. Denis, Reunion

Island, France, February 16–18, 2005; John M.R. Stone, quoted in Bret Schulte, "Special Report: Temperature Rising," *US News & World Report*, June 5, 2006, 39.

56. Philippe Naughton, "Asia-Pacific Climate Pact Takes UK by Surprise," *Times Online*, July 28, 2005. Web address: http://www.timesonline.co.uk/article/0, 2-1711816,00.html; The Asia-Pacific Partnership: Opening Markets for Renewable Energy and Distributed Generation," Conference held Washington, DC, June 6, 2006.
57. Richard Black, "Climate Change Summit Postponed." BBC News, October 5, 2005. Web address: http://news.bbc.co.uk/2/hi/science/nature/4311310.stm.
58. Bill Curry, "Ottawa Now Wants Kyoto Deal Scrapped: Voluntary Climate-change Pact Preferred with Easier Targets, Leaked Paper Reveals," *Globe and Mail*, May 20, 2006. Web address: http:// www.theglobeandmail.com/servlet/story/LAC.20060520.KYOTO20/TPStory/?query=kyoto+leak; also see Gary Park, "Kyoto has Canada in Knots: New Regime Pulls Back from Kyoto Spending; Confusion on Made-in Canada Policy," *Petroleum News*, May 14, 2006. Web address: http://www.petroleumnews.com/pntruncate/691933951.shtml.
59. Frank McDonald, "Climate summit outcome hailed as historic," *Irish Times*, December 12, 2005, 11; Dennis Bueckert, "Canada Decides to go with the Flow on Kyoto Accord: Tories, Who Have Criticized Protocol, Present no Opposition during UN Climate Conference," *The Gazette* (Montreal), May 27, 2006, A3; Aguilar et al. 2005, 14.
60. Curry, "Ottawa Now Wants Kyoto Deal Scrapped."
61. "Mediterranean to Suffer Most in Europe Due to Warming," *Reuters Planet Ark*, October 28, 2005. Web address: http://www.planetark.com/dailynewsstory.cfm/newsid/33199/story.htm.

Bibliography

Adam, David. 2005. 50m Environmental Refugees by End of Decade, UN Warns: States Urged to Prepare for Victims of Climate Change: Natural Disasters Displace More People than Wars. *The Guardian* (October 12): 24.

———. 2006. Hole in Ozone Layer Expected to Increase. *The Guardian* (February 16): 13.

Abbasi, Danial R. 2006. *Americans and Climate Change Closing the Gap between Science and Action: Synthesis of Insights and Recommendations from the 2005 Yale Forestry and Environmental Studies Conference on Climate Change*. New Haven, CT: Yale School of Forestry and Environmental Studies.

Adede, Andronico O. 1995. The Treaty System from Stockholm (1972) to Rio de Janeiro (1992). *Pace Environmental Law Review* 13:33–48.

Agarwal, Anil, Sunita Narain, and Anju Sharma. 1999. *Green Politics: Global Environmental Negotiations*. New Delhi: Centre for Science and Environment.

Agrawala, Shardul and Steinar Andresen. 1999. Indispensibility and Indefensibility? The United States in the Climate Change Negotiations. *Global Governance* 5:457–481.

Aguilar, Soledad, Alexis Conrad, María Gutiérrez, Kati Kulovesi, Miquel Muñoz, and Chris Spence. 2005. Summary of the Eleventh Conference of the Parties to the UN Framework Convention on Climate Change and First Conference of the Parties serving as the Meeting of the Parties to the Kyoto Protocol: November 28– December 10, 2005. *Earth Negotiations Bulletin*, December 12, 2005.

Alexandrowicz, George W. 1996. International Legal Instruments and Institutional Arrangements: A Discussion Paper. In Canadian Council on International Law. *Global Forests & International Environmental Law*. London: Kluwer Law International.

Alliance for Responsible CFC Policy. 1987. The Montreal Protocol: A Briefing Book. Rosslyn, VA: The Alliance for Responsible CFC Policy.

Alvarenga, Karen, Changbo Bai, and Andrey Vavilov. Summary of the Fifteenth Meeting of the Parties to the Montreal Protocol: November 10–14, 2003. *Earth Negotiations Bulletin*, November 17, 2003.

Amerman, Don. 1997. U.S. Producers Seek to Peel Away Barriers; Phytosanitary Bans Still Prove Biggest Hurdle; Smaller Foreign Crops will Help Sales. *Journal of Commerce* (October 17): 7A.

Anderson, J.W. 2001. How the Kyoto Protocol Developed: A Brief History. In *Climate Change Economics and Policy: An RFF Anthology*, ed. Michael Toman. Washington, DC: Resources for the Future.

Association of Small Island States. http://www.sidsnet.org/aosis/.

Axelrod, Robert M. 1984. *The Evolution of Cooperation*. New York: Basic Books.

Baker, Linda. 2000. The Hole in the Sky. *E Magazine: The Environmental Magazine* 11:34–39.

Baldwin, Andrew, Deborah Davenport, Radoslav Dimitrov, Reem Hajjar, and Peter Wood. 2005. Summary of the Fifth Session of the United Nations Forum on Forests: May 16–27, 2005. *Earth Negotiations Bulletin* 13:127.

Baldwin, Paul Ryan. 1999. Innovative Technology, Competitiveness, and Policy Choices at International Environmental Negotiations. PhD diss., Columbia University.

Bang, Guri and Camilla Bretteville Froyn. 2005. U.S. Leadership in Climate Policy—When, Why, and How? Working Paper presented at the International Studies Association Annual Convention, Honolulu, Hawaii, March 1–5, 2005.

Bang, Guri, Andreas Tjernshaugen, and Steinar Andresen. 2005. Future U.S. Climate Policy: International Re-engagement? *International Studies Perspectives* 6:285–303.

Barkin, Samuel and DeSombre, Elizabeth. 2004. The International Community and U.S. Climate Change Policy. Paper presented at the International Studies Association Annual Convention, Montreal, Quebec, Canada, March 3, 2004.

Barkin, J. Samuel and George E. Shambaugh. 1999. Hypotheses on the International Politics of Common Pool Resources. In *Anarchy and the Environment: The International Relations of Common Pool Resources*, ed. J.S. Barkin and George E. Shambaugh. Albany, NY: State University of New York Press.

Barnett, Antony. 2004. Bush Attacks Environment "Scare Stories": Secret Email Gives Advice on Denying Climate Change. *The Observer* (April 4): 22.

Barnsley, Ingrid, Robynne Boyd, Alexis Conrad, and Amber Moreen. 2005. Summary of the Twenty-Fifth Meeting of the Open-Ended Working Group of the Parties to the Montreal Protocol on Substances that Deplete the Ozone Layer and the Second Extraordinary Meeting of the Parties of the Montreal Protocol: June 27–July 1 2005. *Earth Negotiations Bulletin* 19:41.

Barrett, Scott. 2002. Kyoto's Fall. *German-American Relations and the Presidency of George W. Bush*. American Institute for Contemporary German Studies. http://www.aicgs.org/aicgs/aicgs/aicgs/aicgs/aicgs/aicgs/aicgs/aicgs/aicgs/research/g2001/barett.shtml.

———. 2003. *Environment and Statecraft: The Strategy of Environmental Treaty-making*. New York: Oxford University Press.

Barrios, Paula, Alice Bisiaux, Catherine Ganzleben, Amber Moreen, and Chris Spence 2004. Summary of the Sixteenth Meeting of the Parties to the Montreal Protocol: November 22–26, 2004. *Earth Negotiations Bulletin* November 29, 2004.

Barrios, Paula, Noelle Eckley, Pia Kohler, and Dagmar Lohan. 2004. Summary of Extraordinary Meeting of the Parties to the Montreal Protocol: March 24–26, 2004. *Earth Negotiations Bulletin*, March 29, 2004.

Beck, Lindsey. 2005. China Unlikely To Sign On To Kyoto Emissions Cuts. *Macon Daily Online* (November 12). http://www.maconareaonline.com/news.asp?id=12544.

Bellany, Ian. 1997. *The Environment in World Politics: Exploring the Limits*. Cheltenham, UK: Edward Elgar Publishing Ltd.

Benedick, Richard Elliot. 1991. *Ozone Diplomacy: New Directions in Safeguarding the Planet*. Cambridge, MA: Harvard University Press.

Benedick, Richard E. 1993. Perspectives of a Negotiation Practitioner. In *International Environmental Negotiation*, ed. G. Sjöstedt. Newbury Park, CA: Sage Publications.

———. 1997. Do International Environmental Agreements Really Work? The UN Approach to Climate Change: Where Has It Gone Wrong? http://www.weathervane.rff.org/pop/pop4/benedick.html.

———. 1998. *Ozone Diplomacy: New Directions in Safeguarding the Planet*, 2nd ed. Cambridge, MA: Harvard University Press.

Berejekian, Jeffrey. 1997. The Gains Debate: Framing State Choice. *American Political Science Review* 91:789–805.

Berends, Helena. 1993. International Law and Global Environmental Change: A Systematic Study of International Environmental Accords. Paper presented at the European Symposium on National Implementation and Compliance with International Environmental Accord: Blending Academic Perspectives and the Public Sector, June 16–19, 1993, Brussels, Belgium.

Berger, Eric. Keeping its Head Above Water: New Orleans Faces Doomsday Scenario. *The Houston Chronicle* (December 1): A29.

Bernauer, Thomas. 1995. The Effect of International Environmental Institutions: How We Might Learn More. *International Organization* 49:351–377.

Bertrand, Tina L. 1997. Diverging Interests in the Common Good: International Efforts to Regulate the Problem of Acid Rain. Paper presented at the International Studies Association-Southwest Meeting, March 26–29, 1997, New Orleans, LA.

———. 1999. Diverging Interests in the Common Good: International Efforts to Regulate Transboundary Acid Rain. PhD diss., Emory University.

Bettelli, Paola, Chad Carpenter, Deborah Davenport, Peter Doran, and Steve Wise. 1997. The Third Conference of the Parties to the United Nations Framework Convention on Climate Change: December 1–10, 1997. *Earth Negotiations Bulletin*, December 13, 1997.

Black, Richard. 2005. Climate Change Summit Postponed. BBC News (October 5). http://news.bbc.co.uk/2/hi/science/nature/4311310.stm.

———. 2006. Will Kyoto Die at Canadian Hands? BBC News (January 27). http://news.bbc.co.uk/go/pr/fr/-/2/hi/science/nature/4650878.stm.

Blum, Justin. 2006. Making an Oil Pledge; Declaration of Undependence Rests on New Energy Sources. *Washington Post* (February 2): D1.

Bodansky, Daniel. 2001. The History of the Global Climate Change Regime. In *International Relations and Global Climate Change*, ed. Urs Luterbacher and Detlef F. Sprinz, 23–40. Cambridge, MA: MIT Press.

Boehmer-Christiansen, Sonja. 1994. Scientific Uncertainty and Power Politics: The Framework Convention on Climate Change and the Role of Scientific Advice. In *Negotiating Environmental Regimes: Lessons Learned from the United Nations*

Conference on Environment and Development (UNCED), ed. Bertram I. Spector, Gunnar Sjöstedt, and I. William Zartman. London: Graham & Trotman/Martinus Nijhoff.

Botcheva, Liliana and Lisa L. Martin. 2001. Institutional Effects on State Behavior: Convergence and Divergence. *International Studies Quarterly* 45:1–26.

Breidenich, Clare, Daniel Magraw, Anne Rowley, and James W. Rubin. 1998. The Kyoto Protocol to the United Nations Framework Convention on Climate Change. *American Journal of International Law* 92:315–331.

Breitmeier, Helmut. 1997. International Organizations and the Creation of Environmental Regimes. In *Global Governance: Drawing Insights from the Enviornmental Experience*, ed. Oran R. Young, Cambridge, MA: MIT Press.

Broadhead, Lee-Anne. 2002. *International Environmental Politics: The Limits of Green Diplomacy*. Boulder, CO: Lynne Rienner Publishers.

Brown, Lester R. 2000. The Rise and Fall of the Global Climate Coalition. Earth Policy Institute (July 25). http://www.earth-policy.org/Alerts/Alert6.htm

Brown, Michael S. and Katherine A. Lyon. 1992. Holes in the Ozone Layer: A Global Environmental Controversy. In *Controversy: Politics of Technical Decisions*, ed. D. Nelkin. London: Sage.

Brown, Paul. 2005. US Cities Snub Bush and Sign Up to Kyoto: Dozens of Mayors, Representing More Than 29 Million Americans, Pledge to Cut Greenhouse Gases. *Guardian* (May 17): 19.

Brown, Paul and Roger Harrabin. 2005. US Tries to Sink Forests Plan: British Initiative on Illegal Logging Opposed. *The Guardian* (March 16): 15.

Brown, Paul. Rich Nations Could be Sued by Climate Victims. 2001. *The Guardian* (July 10): 7.

Bueckert, Dennis. Canada Decides to go with the Flow on Kyoto Accord: Tories, Who Have Criticized Protocol, Present No Opposition during UN Climate Conference. *The* (Montreal) *Gazette* (May 27): A3.

Bunch, Will. 2005. Did New Orleans Catastrophe Have to Happen? "Times-Picayune" Had Repeatedly Raised Federal Spending Issues. *Editor & Publisher* (August 31), http://www.editorandpublisher.com.

Buncombe, Andrew. 2005. Global Warming: Will You Listen Now, America? *The Independent* (August 19): 1, 3.

Cairncross, Frances. 1992. The Environment: Whose World Is It, Anyway? *The Economist* (May 30): 5–8, 11–18, 21–24.

Caldwell, Lynton Keith with Paul Stanley Weiland. 1996. *International Environmental Policy: From the Twentieth to the Twenty-first Century*. Durham: Duke University Press.

Calvert, Peter. 1994. Environmental Politics and Regime Management. Paper presented at the International Studies Association Annual Convention, Washington, DC, March/April 1994.

Carpenter, Chad, Pamela Chasek, and Steve Wise. 1995. Summary Report on the First Conference of the Parties to the Framework Convention on Climate Change: March 28–April 7, 1995. *Earth Negotiations Bulletin*, April 10, 1995.

Carpenter, Chad, Pamela Chasek, Langston Goree, and Steve Wise. 1995. Summary of the Eleventh Session of the INC for a Framework Convention on Climate Change: February 6–17, 1995. *Earth Negotiations Bulletin*, February 20, 1995.

Carpenter, Chad, Peter Doran, Aarti Gupta, and Lynn Wagner. 1997. Summary of the Nineteenth UN General Assembly Special Session to Review Implementation of Agenda 21: June 23–27. *Earth Negotiations Bulletin*, June 30, 1997.

Carpenter, Chad, Angela Churie, Victoria Kellett, Greg Picker, and Lavanya Rajamani. 1998. Report of the Fourth Conference of the Parties to the UN Framework Convention on Climate Change: November 2–13, 1998. *Earth Negotiations Bulletin,* November 16, 1998.

Carter, Nicole T. 2005. New Orleans Levees and Floodwalls: Hurricane Damage Protection. Congressional Research Service Report RS22238. Washington: Congressional Research Service, The Library of Congress, September 6.

Cass, Loren. 2005. Norm Entrapment and Preference Change: The Evolution of the European Union Position on International Emissions Trading. *Global Environmental Politics* 5:38:60.

Cave, Andrew. 2005. Katrina's Trail of Destruction Spiralling Towards $100bn. *Daily Telegraph* (September 4, 2005): 27.

CEO's to US: Oppose Accelerated HCFC Phaseout. *Engineered Systems* (July 1997): 20.

Change Policy. *Journal of Economic Perspectives* 16:107–130.

Chasek, Pamela S. 1997a. A Comparative Analysis of Multilateral Environmental Negotiations. *Group Decision and Negotiation* 6:437–461.

———. 1997b. The Convention to Combat Desertification: Lessons Learned for Sustainable Development. *Journal of Environment and Development* 6:147–169.

———. 2001. Scientific Uncertainty in Environmental Negotiations. In *Global Environmental Policies,* ed. H.-W. Jeong. London: Palgrave.

Chaterjee, Pratap. 1991. Boycott Could Mar Earth Summit. *New Scientist* (August 24): 72.

Chayes, Abram and Antonia Handler Chayes. 1995. *The New Sovereignty: Compliance With International Regulatory Agreements.* Cambridge, MA: Harvard University Press.

Claussen, Eileen. 1988. Moving Forward Together. *Environmental Forum* (July–August): 14–16.

Clayton, Mark. 2004. New Coal Plants Bury "Kyoto." *Christian Science Monitor* (December 23): 1.

———. 2005. One Region's Bid to Slow Global Warming. *Christian Science Monitor* (December 22): 2.

Climate Change, Clean Energy and Sustainable Development. 2005. Statement of the G8, Gleneagles. http://www.number-10.gov.uk/output/Page7881.asp.

Climate Change Poses Major Threat to U.S. Insurance Industry. 2005. *Insurance Journal* (September 19). http://www.insurancejournal.com/magazines/east/2005/09/19/features/60454.htm.

Climate Convention Negotiations Face Difficulties, but 1992 Target Feasible, Chairman Says. 1991. *United Nations Chronicle* 28 (2):56–57.

Climate Justice: Enforcing Climate Change Law. 2005. http://www.climatelaw.org/.

Climate Signals. 2005. *New York Times* (May 19): 26.

Clover, Charles. 2005. Why the Inuit People are Walking on Thin Ice. *The Daily Telegraph* (October 17): 13.

Collier, Robert. 2005. State Bypasses Kyoto, Fights Global Warming: California Tries to Cut Emissions on its Own. *San Francisco Chronicle* (February 1): A1.

Conference of Parties and Beyond. *EM Online.* http://www.awma.org/em/99/Feb99/features/mathai/mathai.htm.

Connor, Steve. 2005a. Global Warming "Past the Point of No Return." *The Independent* (September 16): 1–2.

———. 2005b. The Final Proof: Global Warming is a Man-made Disaster. *The Independent* (February 19): 1.

Cooper, Mary H. 1991. Saving the Forests. *CQ Researcher* (September 20): 19.

Cooper, Richard. 1992. US Policy Towards the Global Environment. In *The International Politics of the Environment*, ed. A. Hurrell and Ben Kingsbury. Oxford, UK: Clarendon Press.

Corell, Elisabeth. 1997. The Failure of Scientific Expertise to Influence the Desertification Negotiations. Paper presented at the Annual Convention of the International Studies Association, March 18–22, 1997, Toronto, Canada.

Costa Rica-Canada Initiative. 1999. Final Meeting. December 6–10. http://www.iisd.ca/sd/crci/final/.

Costello, Matthew J. 1996. Impure Public Goods, Relative Gains, and International Cooperation. *Policy Studies Journal* 24:578–594.

Curry, Bill. 2006. Ottawa Now Wants Kyoto Deal Scrapped: Voluntary Climate-Change Pact Preferred with Easier Targets, Leaked Paper Reveals. *Globe and Mail* (May 20). http://www.theglobeandmail.com/servlet/story/LAC.20060520.KYOTO20/TPStory/?query=kyoto+leak.

Cushman, John H. 1998. Big Problem, Big Problems: Getting to Work on Global Warming. *New York Times* (December 8): G4.

Dales, J.H. 1992. The Property Interface. In *Environmental Economics: A Reader*, ed. Anil Markandya and Julie Richardson. New York: St. Martin's Press.

Daly, Herman. 1982. Three Visions of the Economic Process. In *International Dimensions of the Environmental Crisis*, ed. R.N. Barrett. Boulder, CO: Westview Press.

———. 1991. Elements of Environmental Macroeconomics. In *Ecological Economics*, ed. Robert Costanza. New York: Columbia University Press.

Davenport, Deborah. 2005. An Alternative Explanation for the Failure of the UNCED Forest Negotiations. *Global Environmental Politics* 5:105–130.

Davenport, Deborah, Laura Ivers, Leila Mead, and Kira Schmidt. 1998. Second Session of the Intergovernmental Forum on Forests, August 24–September 4, 1998. *Earth Negotiations Bulletin*, September 7, 1998.

Davenport, Deborah, Nabiha Megateli, Kira Schmidt, and Steve Wise. 1997. Summary of the Fourth Session of the Intergovernmental Panel on Forests: February 11–21, 1997. *Earth Negotiations Bulletin*, February 24, 1997.

Deen, Thalif. 1992a. Desertificacion y Bosques: EEUU Esperaba un Canje? *Terra Viva* (June 12): 9.

———. 1992b. Mecanismos de Financiamiento Dividieron al Grupo de los 77. *Terra Viva* (June 9): 4.

Demetrio, Patricia B. 1997. Unfair to US Farms. *Journal of Commerce* (December 18): 1A.

Denemark, Robert A. 1999. World Systems History: From Traditional International Politics to the Study of Global Relations. *International Studies Review* 1:43–75.

Depledge, Joanna, Ian W. Fry, Laura Ivers, and Chris Spence. 1999. Summary of the Eleventh Meeting of the Parties to the Montreal Protocol and the Fifth Conference of the Parties to the Vienna Convention: November 29—December 3, 1999. *Earth Negotiations Bulletin*, December 6, 1999.

Depledge, Joanna, Laura Ivers, Herman Lopez, Lavanya Rajamani, and Lisa Schipper. 2000. Summary of the Twelfth Meeting of the Parties to the Montreal Protocol on Substances that Deplete the Ozone Layer: December 11–14, 2000. *Earth Negotiations Bulletin*, December 15, 2000.

Depledge, Joanna, Laura Ivers, Herman Lopez, and Thomas Yongo. 2000. Summary of the Twentieth Meeting of the Open-Ended Working Group of the Parties to the Montreal Protocol on Substances that Deplete the Ozone Layer: July 11–13, 2000. *Earth Negotiations Bulletin*, July 14, 2000.

Depledge, Joanna, Andrei Henry, Laura Ivers, and Kira Schmidt. Summary of the Thirteenth Meeting of the Parties to the Montreal Protocol on Substances that Deplete the Ozone Layer: 16-October 19, 2001. *Earth Negotiations Bulletin*, October 22, 2001.

DeSilver, Drew. 2001. Tariff Impact Not Clear-cut: Weyerhaeuser Has Stake in Lumber Dispute. *Seattle Times* (August 22): C1.

DeSombre, Elizabeth R. 2000. *Domestic Sources of International Environmental Policy: Industry, Environmentalists, and U.S. Power*. Cambridge, MA: MIT Press.

———. 2002. *The Global Environment and World Politics*. New York: Continuum.

DeSombre, Elizabeth R. and Joanne Kauffman. 1996. The Montreal Protocol Multilateral Fund: Partial Success Story. In *Institutions for Environmental Aid: Pitfalls and Promise*, ed. Robert O. Keohane and Marc A. Levy. Cambridge, MA: MIT Press.

Dimitrov, Radoslav S. 2003. Knowledge, Power, and Interests in Environmental Regime Formation. *International Studies Quarterly* 47:123–150.

Dimitrov, Radoslav, Laura Ivers, Leila Mead, and Kira Schmidt. 2001. Summary of the First Session of the United Nations Forum on Forests, June 11–23. *Earth Negotiations Bulletin* 13:83.

Division of Global Change, Office of Air and Radiation. 1989. Costs of Controlling Methyl Chloroform in the US (Review Draft). Washington, DC: US Environmental Protection Agency.

Doniger, David. 1988. Politics of the Ozone Layer. *Issues in Science and Technology* 4:86–92.

Dotinga, Harm. 1998. 52nd Session: Reform, Environment, and Development-related Issues. *Environmental Policy and Law* 28:21–30.

Dotto, Lydia and Harold Schiff. 1978. *The Ozone War*. Garden City, NY: Doubleday.

Downie, David Leonard. 1996. Understanding International Environmental Regimes: Lessons of the Ozone Regime. PhD Diss., University of North Carolina.

———. 1999. The Power to Destroy: Understanding Stratospheric Ozone Politics as a Common-Pool Resource Problem. In *Anarchy and the Environment: The International Relations of Common Pool Resources*, ed. S.J. Barkin and George E. Shambaugh. Albany, NY: State University of New York Press.

Downs, Anthony. 1972. Up and Down With Ecology: The "Issue-Attention Cycle." *The Public Interest* 28:38–50.

Downs, George W., David M. Rocke, and Peter N. Barsoom. 1996. Is the Good News About Compliance Good News About Cooperation? *International Organization* 50:379–406.

———. 1997. Designing Multilaterals: The Architecture of Environmental Agreements. Paper presented at the American Political Science Association Annual Conference, Washington, DC, August 14, 1997.

———. 1998. Managing the Evolution of Multilateralism. *International Organization* 52:397–419.

Downs, George W., David M. Rocke, and Randolph M. Siverson. 1986. Arms Races and Cooperation. In *Cooperation Under Anarchy*, ed. K.A. Oye. Princeton, NJ: Princeton University Press.

Draft of Environmental Rules: "Global Partnership". 1992. *New York Times* (April 5): A6.

Dunne, Nancy. 2001. Bush Backs Away from Pledge to Curb Carbon Emissions. *Financial Times* (March 14): 1.

Du Pont Gets OK to Market Blowing Agent. 1988. *Journal of Commerce* (January 8): 9B.

Du Pont Position Statement on the Chlorofluorocarbon/Ozone/Greenhouse Issues. 1997. *Environmental Conservation* 13:363–364.

Du Pont Produces CFC Replacement on Small Scale. 1988. *Journal of Commerce* (February 3): 9B.

Du Pont Sends a Message on Ozone. 1988. *New York Times* (March 29): A26.

Dudek, Daniel J. and Michael Oppenheimer. 1986. The Implications of Health and Environmental Effects for Policy. In *Effects of Changes in Stratospheric Ozone and Global Climate*, ed. J.G. Titus. Washington, DC: USEPA and UNEP.

Eberstadt, Nicholas. 1997. The UN's Development Activities. *World Affairs* 159:151–157.

Eilperin, Juliet. 2006. Debate on Climate Shifts to Issue of Irreparable Change; Some Experts on Global Warming Foresee "Tipping Point" When It Is Too Late To Act. *Washington Post* (January 29): A01.

Elkins, James W. 1999. Chlorofluorocarbons (CFCs). In *The Chapman & Hall Encyclopedia of Environmental Science*, ed. David E. Alexander and Rhodes W. Fairbridge, 78–80. Boston, MA: Kluwer Academic.

Elliot, Joy. 1992a. Jury Undecided on GEF's First Year. *Earth Summit Times* (June 2): 3.

———. 1992b. Nations Agree to Draft Desert Convention. *Earth Summit Times* (June 11): 2.

Elliott, Lorraine. 1998. *The Global Politics of the Environment*. London: MacMillan Press, Ltd.

Elster, Jon. 1986. *Rational Choice, Readings in Social and Political Theory*. Oxford: Basil Blackwell.

Environmental Resources Ltd. 1979. Protection of the Ozone Layer—Some Economic and Social Implications of a Possible Ban on the Use of Fluorocarbons. In *The Ozone Layer: Proceedings of the Meeting of Experts Designated by Governments, Intergovernmental and Nongovernmental Organizations on the Ozone Layer Organized by the United Nations Environment Programme in Washington D.C., March 1–9, 1977*, ed. A.K. Biswas. Oxford: Pergamon Press, pp. 141–175.

Examination of International Forestry Regulation, Both Public and Private. *UCLA Journal of Environmental Law & Policy* 19:153–180.

Executive Office of the President. 1990. White House Press Release. Elements of a Forest Convention. Washington, DC.

———. 1992. White House Press Release. Bush Proposes "Forests for the Future" Initiative. Rio de Janeiro, Brazil.

Fairley, Peter. 1996. Clinton Pressured to Ease 2001 Ban. *Chemical Week* (October 16): 15.

Farman, J.C. et al. 1985. Large Losses of Total Ozone in Antarctica Reveal Seasonal CLO_x/NO_x Interactions. *Nature* (May 16): 207–210.

Faucheux, Sylvie and Geraldine Froger. 1995. Decision-Making Under Environmental Uncertainty. *Ecological Economics* 15:29–42.

Faure, Guy-Olivier and Jeffrey Z. Rubin. 1993. Original Concepts and Questions. In *International Environmental Negotiation*, ed. G. Sjöstedt. Newbury Park, CA: Sage Publications.

Fearon, James D. 1991. Counterfactuals and Hypothesis Testing in Political Science. *World Politics* 43:169–195.

———. 1994. Domestic Political Audiences and the Escalation of International Disputes. *American Political Science Review* 88:577–592.

———. 1998. Bargaining, Enforcement, and International Cooperation. *International Organization* 52:269–305.

Feder, Barnaby J. 2004. A Different Era for the Alternative Energy Business. *New York Times* (May 29): C1.

Federovisky, Sergio. 1992a. Crece el Optimismo: EEUU Daría Marcha Atrás y Firmaría. *Terra Viva* (June 14): 11.

———. 1992b. EEUU Quiere Reabrir el Debate. *Terra Viva* (June 3): 2.

Feldman, David L. 1990. Managing Global Climate Change through International Cooperation: Lessons From Prior Resource Management Efforts. Oak Ridge, TN: Oak Ridge National Laboratory Global Environmental Studies Center.

Fierke, K.M. and Michael Nicholson. 1999. Divided by a Common Language: Formal and Constructivist Approaches to Games. Paper presented at the International Studies Association Annual Conference, Washington, DC, February 16–20, 1999.

Fisher, Roger, William Ury, and Bruce Patton. 1991. *Getting to Yes: Negotiating Agreement Without Giving In*. Boston: Houghton Mifflin.

Florini, Ann. 1996. The Evolution of International Norms. *International Studies Quarterly* 40:363–389.

Frank, Robert H. 1996. What Price the Moral High Ground? *Southern Economics Journal* 63:1–17.

Frank, Robert H., Thomas Gilovich, and Dennis T. Regan. 1993. Does Studying Economics Inhibit Cooperation? *Journal of Economic Perspectives* 7:159–171.

———. 1998. Money Well Spent? In *Never Enough: The New Conspicuous Consumption*, ed. R.H. Frank. New York: The Free Press.

French, Hilary. 1997. Learning from the Ozone Experience. In *State of the World, 1997* ed. Cester R. Brown. New York: W.W. Norton.

Fulmer, Melinda. 2001. Strawberry Fields May Not Be Forever; Agriculture: Phaseout as a Key Fumigant Has California's Growers Bracing for Trouble over the Next Few Years. Los Angeles *Times* (January 11): C1.

G-7, Developing Country Communiqués Highlight UNCED Conflicts. 1991. *Earth Summit Update* (August): 2.

Gale, Fred. 1998. *The Tropical Timber Trade Regime*. Basingstoke, Hampshire: Macmillan Press.

GE Launches Ecomagination to Develop Environmental Technologies; Company-Wide Focus on Addressing Pressing Challenges. 2005. General Electric Company Press Release (May 9).

Gerson, Noel L. 1992. North-South Agreement on Forests. *Earth Summit Times* (June 13): 9.

Gerstenzang, James. 1992. Bush Proposes Huge Growth in Forest Funding. *Los Angeles Times* (June 2): A10.

Ghazi, Polly. 2005. A Storm Brewing: Is Bush Out of Step with US Public Opinion? *The Guardian* (July 6): 13.

Gilpin, Robert. 1975. *U.S. Power and the Multinational Corporation: The Political Economy of Foreign Direct Investment*. New York: Basic Books.

———. 1981. *War and Change in World Politics*. Cambridge, NY: Cambridge University Press.

———. 1987. *The Political Economy of International Relations*. Princeton, NJ: Princeton University Press.

Girard, Daniel. 2001. At Loggerheads. *Toronto Star* (August 25): 3.

Gladwell, Malcolm. 1988. Du Pont Plans to Make CFC Alternative. *Washington Post* (September 30): F5.

Global Warming Follies. 2002. *New York Times* (June 8): A14.

Goodland, Robert, Herman Daly, and Salah el Serafy. 1991. *Environmentally Sustainable Economic Development: Building on Brundtland*. Washington, DC: World Bank.

Goodstein, Laurie. 2006. Evangelical Leaders Join Global Warming Initiative. *New York Times* (February 8): A12.

Gore, Senator Albert and Congressman John Porter. 1991. Joint Resolution Calling on the President of the United States to Take a Leadership Role in the International Negotiations Toward a World Forest Convention and a Framework Convention on Climate Change, and for Other Purposes (S.J. Res. 181/H.J. Res. 302.) Washington, DC.

Gram, David. 2005. Vermont Adopts New Rules to Cut Car CO2 Emissions: New Standards Take Effect for 2009 Model Year. *Detroit News Autos Insider* (November 3). http://www.detnews.com/2005/autosinsider/0511/03/0auto-370678.htm.

Grieco, Joseph. 1988. Anarchy and the Limits of Cooperation: A Realist Critique of the Newest Liberal Institutionalism. *International Organization* 42:485–586.

Group of 77 website: www.g77.org/main/main.htm.

Grubb, Michael, Matthias Koch, Koy Thomson, Abby Munson, and Francis Sullivan. 1993. *The Earth Summit Agreements: A Guide and Assessment, an Analysis of the Rio '92 UN Conference on Environment and Development.* London: Earthscan Publications Ltd.

Grundmann, Reiner. 1998. The Strange Success of the Montreal Protocol: Why Reductionist Accounts Fail. *International Environmental Affairs* 10:197–220.

Guppy, Nicholas. 1996. International Governance and Regimes Dealing with Land Resources from the Perspective of the North. In *Global Environmental Change and International Governance*, ed. Oran Young, George K. Demko, and Kilaparti Ramakrishna. Providence, RI: University Press of New England.

Haas, Peter M. 1989. Do Regimes Matter? Epistemic Communities and Mediterranean Pollution Control. *International Organization* 43:377–403.

———. 1992a. Banning Chlorofluorocarbons: Efforts to Protect Stratospheric Ozone. *International Organization* 46:187–224.

———. 1992b. Climate Change Negotiations. *Environment* 34:2–3.

———. 1992c. Epistemic Communities and International Policy Coordination. *International Organization* 46:1–36.

———. 1993. Stratospheric Ozone: Regime Formation in Stages. In *Polar Politics: Creating International Environmental Regimes*, ed. Oran R. Young and Gail Osherenko. Ithaca, NY: Cornell University Press.

Haas, Peter M., Robert O. Keohane, and Marc A. Levy. 1993. *Institutions for the Earth: Sources of Effective International Environmental Protection*. Cambridge, MA: MIT Press.

Haggard, Stephan and Beth A. Simmons. 1987. Theories of International Regimes. *International Organization* 41:491–517.

Haigh, Nigel. 1989. *EEC Environmental Policy and Britain*. London: Longman.

———. 1992. The European Community and International Environmental Policy. In *The International Politics of the Environment*, ed. Andrew Hurrell and Benedict Kingsbury. Oxford, UK: Clarendon Press.

Hajjar, Reem, Twig Johnson, Harry Jonas, Leila Mead, and Peter Wood. 2006. Summary of the Sixth Session of the United Nations Forum on Forests: February 13–24, 2006. *Earth Negotiations Bulletin*, February 27, 2006.

Hampson, Fen Osler and Michael Hart. 1995. *Multilateral Negotiations: Lessons from Arms Control, Trade, and the Environment*. Baltimore, MD: Johns Hopkins University Press.

Hansen, Jim. 2006. Greenland Ice Cap Melting at Twice the Rate It Was Five Years Ago, Says Scientist Bush Tried to Gag. *Independent* (February 17): 1.

Hardin, Garrett. 1968. The Tragedy of the Commons. *Science* 162:1243–1248.

Hardin, Russell. 1982. *Collective Action*. Baltimore, MD: Johns Hopkins University Press.

Harkavy, Angela. 1992. *A Progress Report on Preparatory Negotiations for the United Nations Conference on Environment and Development: The Final Effort.* Rio de Janeiro: National Wildlife Federation and CAPE '92.

Harrabin, Roger. 2005. Bush in Climate Change Rebellion. BBC News (November 9). http://newssearch.bbc.co.uk/cgibin/search/results.pl?scope=newsifs&tab=news&q=climate+change.

Harris, Paul G. 2001. International Environmental Affairs and U.S. Foreign Policy. In *The Environment, International Relations, and US Foreign Policy*, ed. Paul Harris. Washington, DC: Georgetown University Press.

Harsanyi, John C. 1986. Advances in Understanding Rational Behavior. In *Rational Choice*, ed. J. Elster. New York: New York University Press.

Harvey, Fiona. 2006. Beijing Finds Love in a Changing Climate. *Financial Times* (January 27): 17.

Hasenclever, Andreas, Peter Mayer and Volker Rittberger. 1996. Interests, Power, Knowledge: The Study of International Regimes. *Mershon International Studies Review* 40:177–228.

Heap, Shaun Hargreaves. 1992. *The Theory of Choice: A Critical Guide*. Oxford, UK: Blackwell.

Heissenbuttel, John, Charlotte Fox, Gerald Gray, and Gary Larsen. 1992. *Principles for Sustainable Management of Global Forests: Review of the Forest Principles and Agenda 21, Chapter 11, "Combating Deforestation."* Washington, DC: The Global Forestry Coordination and Cooperation Project.

Hempel, Lamont C. 1996. *Environmental Governance: The Global Challenge.* Washington, DC: Island Press.

Hicks, Jonathan P. 1987. Chemical Industry Sees Rush to Invent Safer Alternatives. *New York Times* (September 17): A12.

Higgins, Andrew. 1992. The Earth Summit: China Seeks Funds for Fight against Black Rain. *The Independent* (May 25): 8.

History of the Montreal Protocol's Ozone Fund. *International Environmental Reporter*, (November 20): 636–640.

Hoesung Lee, Bruce J. and E. Haites, eds. 1996. *Climate Change 1995: Economic and Social Dimensions of Climate Change, Contribution of Working Group III to the Second Assessment Report of the Intergovernmental Panel on Climate Change.* Cambridge, UK: Cambridge University Press.

Holdgate, Martin W. 1991. The Environment of Tomorrow. *Environment* 33:14–20, 40–42.

Hole in Arctic Ozone is Feared. 1988. *New York Times* (May 18): A25.

Holmberg, Johan. 1992. Financing Sustainable Development. In *Making Development Sustainable: Redefining Institutions, Policy, and Economics*, ed. J. Holmberg. Washington, DC: Island Press.

Holusha, John. 1990. Du Pont to Construct Plants for Ozone-Safe Refrigerant. *New York Times* (June 23): 31.

———. 2006. California Bill Calls for Cuts in Emissions. *New York Times* (April 4): A19.

Hook, Steven W. 1994. Self-Interest and Foreign Economic Policy: A Cross-National Perspective. *International Studies Notes* 19:26–36.

Houlder, Vanessa. 2003. Climate Change Could Be Next Legal Battlefield. *Financial Times* (July 14): 10.
Houston Economic Declaration: Houston Economic Summit, July 11, 1990. http://bushlibrary.tamu.edu/research/papers/1990/90011100.htm.
Hughes, Elaine L. 1996. Forests, Forestry Practices and the Living Environment. In Canadian Council on International Law. *Global Forests & International Environmental Law*, London: Kluwer Law International.
Hughes, J. Donald. 1975. *Ecology in Ancient Civilizations*. Albuquerque, NM: University of New Mexico Press.
Hughes, Sylvia. 1991. North-South Split Kills Hopes of Forest Treaty. *New Scientist* (September 28): 14.
Humphreys, David. 1993. The Forests Debate of the UNCED Process. *Paradigms* 7:43–54.
———. 1996a. *Forest Politics: The Evolution of International Cooperation*. London: Earthscan Publications Ltd.
———. 1996b. The Global Politics of Forest Conservation since the UNCED. *Environmental Politics* 5:231–256.
———. 2001. The Creation of the United Nations Forum on Forests. *Environmental Politics* 10:160–166.
———. 2005. The Elusive Quest for a Global Forests Convention. *Review of European Community and International Environmental Law (RECIEL)* 14:1–10.
Hurrell, Andrew, and Benedict Kingsbury. 1992. *The International Politics of the Environment: Actors, Interests, and Institutions*. New York: Clarendon Press.
Hurtado, Maria Elena. 1992. Northern Obsession with Forests Boggles the Mind. *Crosscurrents* (June 5): 12.
Ignatius, David. 2004. Dishonest Trade Talk. *Washington Post* (February 24): A21.
Imber, Mark. 1991. The UN Environment Agenda. Paper presented at the British International Studies Association: Global Environmental Change Group Meeting, February 22, 1991, London.
Immelt, Jeffrey. 2005. Ecomagination. Statement made May 9, 2005.
Insurers Link Global Warming with Higher Cost of Storms. 2005. *Environment News Service* (July 28). http://www.ens-newswire.com/ens/jul2005/2005-07-25-02.asp.
Issue-by-Issue Summary of PrepCom 3. 1991. *Earth Summit Update* (October): 1–6.
Intergovernmental Panel on Climate Change. 2001. Third Assessment Report—Climate Change 2001. WMO/UNEP. http://www.ipcc.ch/pub/reports.htm.
Izzo, Dominic. 2004. Reengineering the Mississippi. *Civil Engineering Magazine* (July). http://www.pubs.asce.org/ceonline/ceonline04/0704feat.html.
Jachtenfuchs, Markus. 1990. The European Community and the Protection of the Ozone Layer. *Journal of Common Market Studies* 28:261–278.
Johnson, Stanley. 1993. *The Earth Summit: The United Nations Conference on Environment and Development (UNCED)*. London: Graham & Trotman/Martinus Nijhoff.
Jurgielewicz, Lynne M. 1996. *Global Environmental Change and International Law: Prospects for Progress in the Legal Order*. Lanham, MD: University Press of America.

Justus, John R. and Susan R. Fletcher. 2001. CRS Issue Brief for Congress. IB89005: Global Climate Change. Congressional Research Service. August 13. http://www.ncseonline.org/NLE/CRSreports/Climate/clim-2.cfm?&CFID=1732361&CFTOKEN=90447656#_1_12.

Kaelberer, Matthias. 1992. State Power and International Environmental Cooperation: A Comparison of Collaborative Efforts on Global Warming and Ozone Layer Depletion. Paper presented at the International Studies Association Annual Conference, Atlanta, Georgia, March 31-April 4, 1992.

Kahneman, Daniel. 2002. Maps of Bounded Rationality: A Perspective on Intuitive Judgment and Choice. http://nobelprize.org/economics/laureates/2002/kahnemann-lecture.pdf.

Kapstein, Ethan B. 2000. Winners and Losers in the Global Economy. *International Organization* 54:259–384.

Kay, Jane. 2004. State Seeks 30% Cut in Tailpipe Emissions; Automakers Say They May Sue If 10-Year Plan OKd. *San Francisco Chronicle* (June 15): A1.

Keohane, Robert O. 1984. *After Hegemony: Cooperation and Discord in the World Political Economy*. Princeton, NJ: Princeton University Press.

———. 1997. Problematic Lucidity: Stephen Krasner's "State Power and the Structure of International Trade." *World Politics* 50:150–170.

———. 2005. *After Hegemony: Cooperation and Discord in the World Political Economy*. Princeton: Princeton University Press.

King, Gary, Robert O. Keohane, and Sidney Verba. 1994. *Designing Social Inquiry: Scientific Inference in Qualitative Research*. Princeton, NJ: Princeton University Press.

Kiss, Alexandre Charles and Dinah Shelton. 2000. *International Environmental Law*. Ardsley, NY: Transactional Publishers.

Knight, Jack. 1992. *Institutions and Social Conflict*. Cambridge, UK: Cambridge University Press.

Kolk, Ans. 1996a. Forest Issues in the International Arena: From UNCHE to UNCED and Beyond. In *International Environmental Law & Policy Occasional Paper Series, No. 12*. Berkeley, CA: The GreenLife Society.

———. 1996b. *Forests in International Environmental Politics: International Organisations, NGOs and the Brazilian Amazon*. Utrecht, The Netherlands: International Books.

Koremenos, Barbara, Charles Lipson, and Duncan Snidal. 2001. The Rational Design of International Institutions. *International Organization* 55:761–799.

Kottary, Sailesh. 1992. Indian Minister Criticizes Forest Text. *Earth Summit Times* (June 9): 15.

Krasner, Stephen D. 1976. State Power and the Structure of International Trade. *World Politics* 28:317–347.

———. 1983. Structural Causes and Regime Consequences: Regimes as Intervening Variables. In *International Regimes*, ed. Stephen D. Krasner. Ithaca, NY: Cornell University Press.

———. 1985. *Structural Conflict: The Third World against Global Liberalism*. Berkeley, CA: University of California Press.

———. 1991. Global Communications and National Power: Life Along the Pareto Frontier. *World Politics* 43:336–366.

Kremenyuk, Viktor A. and Guy Faure. 1991. *International Negotiation: Analysis, Approaches, Issues*. San Francisco: Jossey-Bass Publishers.

Kremenyuk, Viktor A. and Winfried Lang. 1993. The Political, Diplomatic, and Legal Background. In *International Environmental Negotiation*, ed. G. Sjöstedt. Newbury Park, CA: Sage Publications.

Kriz, Margaret. 2005a. Fueling the Dragon. *National Journal* (August 6): 2510–2513.

———. 2005b. Heating Up: Global Warming Moves to a Front Burner, as Demands Grow for Aggressive Action to Limit Greenhouse-Gas Emissions. *National Journal* (August 6): 2504–2508.

Kütting, Gabriela. 2000. *Environment, Society and International Relations: Towards More Effective International Environmental Agreements*. London: Routledge.

Kyoto is Dead Duck Until US Clambers on Board. 2005. *Insurance Day* (February 16). http://www.insuranceday.com/insday/homepage.jsp?pageid=article&articleid=20000068436.

Lamborn, Alan C. 1997. Theory and the Politics in World Politics. *International Studies Quarterly* 41:187–214.

Lammers, Johan G. 1988. Efforts to Develop a Protocol on Chlorofluorocarbons to the Vienna Convention for the Protection of the Ozone Layer. In *Hague Yearbook of International Law 1988*. The Hague, Netherlands: Martinus Nijhoff.

Lane, Earl. 1992. Proposal Doubles Money for Forests. *Newsday* (June 2): 79.

Lang, Winfried. 1997. *The Ozone Treaties and their Influence on the Building of International Environmental Regimes*. Vienna, Austria: Bundesministerium für Auswärtige Angelegenheiten.

Lawton, Valerie. 2001. Canada Stays Calm on Kyoto Pullout. *The Toronto Star* (March 30).

Lax, David A. and James K. Sebenius. 1986. *The Manager as Negotiator: Bargaining for Cooperative and Competitive Gains*. New York: The Free Press.

Lean, Geoffrey. 2004. Global Warming Will Plunge Britain into New Ice Age "Within Decades." *The Independent* (January 25): 18.

Levick, Diane. 2005. Insurers Tackle Climate; Industry Urged to Seek New Types of Coverage. *The Hartford Courant* (October 28): E1.

Levy, Jack S. 1997. Prospect Theory, Rational Choice, and International Relations. *International Studies Quarterly* 41:87–112.

Lewis, Damien and John Vidal. 1990. Environment: UN Proposes Full Forestry Convention. *The Guardian* (June 15).

Lewis, Paul. 1992a. Balancing Industry with Ecology. *New York Times* (March 2): A6.

———. 1992b. Environment Aid for Poor Nations Agreed at the U.N. *New York Times* (April 5): A1.

———. 1992c. U.S. Under Fire in Talks at UN on Environment. *New York Times* (March 24): A1.

Lijphart, Arend. 1971. Comparative Politics and the Comparative Method. *American Political Science Review* 65:682–693.

Lipschutz, Ronnie D. 2000/2001. Why Is There No International Forestry Law?: An Examination of International Forestry Regulation, Both Public and Private. *UCLA Journal of Environmental Law & Policy* 19:153–180.

———. 2004. *Global Environmental Politics: Power, Perspectives, and Practice.* Washington, DC: CQ Press.

List, Martin and Volker Rittberger. 1992. Regime Theory and International Environmental Management. In *The International Politics of the Environment*, ed. Andrew Hurrell and Benedict Kingsbury. Oxford, UK: Clarendon Press.

Litfin, Karen T. 1994. *Ozone Discourses: Science and Politics in Global Environmental Cooperation.* New York: Columbia University Press.

———. 1995. Framing Science: Precautionary Discourse and the Ozone Treaties. *Millennium: Journal of International Studies* 24:251–277.

———. 1996a. The Politics of Global Atmospheric Change. In *International Studies Quarterly* 40:279–283.

———. 1996b. Up in the Air: Dynamic Regimes for the Global Atmosphere. *Mershon International Studies Review* 40:279–283.

———. 1997. Sovereignty in World Ecopolitics. *Mershon International Studies Review* 41:167–204.

Lomborg, Bjørn. 2001. *The Skeptical Environmentalist: Measuring the Real State of the World.* Cambridge, UK: Cambridge University Press.

Maassarani, Tarek. 2006. Muzzling scientists on warming. *Topeka Capital-Journal* (May 19). Web address: http://www.cjoline.com/stories/051906/opi_massarani.shtml.

Maathai, C.V. and Mark C. Trexler. 1999. Global Climate Change: The Buenos Aires Conference of Parties and Beyond. *EM Online.* http://www.awma.org/em/99/Feb99/features/mathai/mathai.htm.

Mankin, William E. 1998. Entering the Fray: International Forest Policy Processes: An NGO Perspective on their Effectiveness (Discussion Paper). London: International Institute for Environment and Development.

Martin, Lisa L. 1992a. *Coercive Cooperation: Explaining Multilateral Economic Sanctions.* Princeton, NJ: Princeton University Press.

———. 1992b. Interests, Power, and Multilateralism. *International Organization* 46:765–792.

Mathews, Jessica Tuchman. 1991. *Preserving the Global Environment: The Challenge of Shared Leadership.* New York: W.W. Norton.

Maxwell, James H. and Sanford L. Weiner. 1993. Green Consciousness or Dollar Diplomacy? The British Response to the Threat of Ozone Depletion. *International Environmental Affairs* 5:19–41.

McCarthy, Michael. 1992. Counting the Trees Brings Confusion. *The Times* (London) (June 2): 10.

———. 2005. Climate Change Poses Threat to Food Supply, Scientists Say. *The Independent* (April 27): 11.

McDonald, Frank. 2005. Climate Summit Outcome Hailed as Historic. *Irish Times* (December 12): 11.

McGregor, Alan. 1987. Deal Near on Saving World Ozone Layer. *The Times* (April 30): 10.

McKibbin, Warwick J. and Peter J. Wilcoxen. 2002. The Role of Economics in Climate Change Policy. *Journal of Economic Perspectives* 16:107–129.

McMurray, Scott. 1993. Du Pont to Speed Up Phaseout of CFCs as Ozone Readings Post Record Lows. *Wall Street Journal* (March 9): B5.

Mediterranean to Suffer Most in Europe Due to Warming. 2005. *Reuters Planet Ark* (October 28). http://www.planetark.com/dailynewsstory.cfm/newsid/33199/story.htm.

Metzger, Jennifer. 2005. Expertise and Corporate Power: Economic Actors and the Stratospheric Ozone Policy Debate. Paper presented at the annual meeting of the International Studies Association, Hilton Hawaiian Village, Honolulu, Hawaii, March 5, 2005.

Mickelson, Karin. 1996. Seeing the Forest, the Trees and the People: Coming to Terms with Developing Country Perspectives on the Proposed Global Forests Convention. In Canadian Council on International Law. *Global Forests and International Environmental Law*. London: Kluwer Law International.

Mikesell, Raymond F. 1999. Review Article on Climate of Fear: Why We Shouldn't Worry about Global Warming by Thomas Gale Moore. *Journal of Economic Literature* 37:1212–1213.

Miles, Edward L. 2002. *Environmental Regime Effectiveness: Confronting Theory with Evidence*. Cambridge, MA: MIT Press.

Miller, Alan S. 1989. Incentives for CFC Substitutes: Lessons for Other Greenhouse Gases. In *Coping with Climate Change*, ed. J.D. Topping (Proceedings of the Second North American Conference on Preparing for Climate Change: A Cooperative Approach, held December 6–8, 1988, in Washington, DC).

Miller, Alan S. and Irving M. Mintzer. 1986. *The Sky is the Limit: Strategies for Protecting the Ozone Layer*. Washington, DC: The World Resources Institute.

Miller, Kenton and Laura Tangley. 1991. *Trees of Life: Saving Tropical Forests and their Biological Wealth*. Boston, MA: Beacon Press.

Miller, Marian A.L. 1995. *The Third World in Global Environmental Politics*. Boulder, CO: Lynne Rienner Publishers.

Mitchell, Ronald B. 1994. Regime Design Matters: Intentional Oil Pollution and Treaty Compliance. *International Organization* 48:425–458.

———. 1999a. International Environmental Common Pool Resources: More Common than Domestic but More Difficult to Manage. In *Anarchy and the Environment: The International Relations of Common Pool Resources*, ed. J.S. Barkin and George E. Shambaugh. Albany, NY: State University of New York Press.

———. 1999b. Situation Structure and Regime Implementation Strategies. Paper presented at the American Political Science Annual Conference, Atlanta, GA, September, 1999.

Modelski, George and Kazimierz Poznanski. 1996. Evolutionary Paradigms in the Social Sciences. *International Studies Quarterly* 40:315–319.

Molina, Mario and F. Sherwood Rowland. 1974. Stratospheric Sink for Chlorofluoromethanes: Chlorine Atom-Catalysed Destruction of Ozone. *Nature* (June 28): 810–812.

Monastersky, Richard. 1990a. Bush Holds Cautious Course on Global Change. *Science News* 137 (7): 102.

Monastersky, Richard. 1990b. Bush Holds Cautious Course on Global Change. *Science News* 137 (17): 263.

Mongelluzzo, Bill. 1998. Pesticide Ruling Will Hurt Exporters. *Journal of Commerce* (December 19): 1A.

Montreal Protocol Technical and Economic Assessment Panel. HCFC Task Force Report. May 2003. http://www.fluorocarbons.org/documents/library/TEAP_HCFC_2003.pdf.

Moran, Emilio, Elinor Ostrom, and J.C. Randolph. 1998. A Multilevel Approach to Studying Global Environmental Change in Forest Ecosystems. Paper presented at the Fifth Biennial Meeting of the International Society for Ecological Economics, November 15–19, 1998, Santiago, Chile.

Moravcsik, Andrew. 1999. A New Statecraft? Supranational Entrepreneurs and International Cooperation. *International Organization* 53:267–306.

Moreira, Naila. 2006. Ozone "Recovery" May be Solar Trick. *ScienceNOW Daily News* (February 13). http://sciencenow.sciencemag.org/cgi/content/full/2006/213/1?eaf.

Morgan, T. Clifton. 1994. *Untying the Knot of War: A Bargaining Theory of International Crises*. Ann Arbor, MI: University of Michigan Press.

Morrisette, Peter M. 1989. The Evolution of Policy Responses to Stratospheric Ozone Depletion. *Natural Resources Journal* 29:793–820.

———. 1991. The Montreal Protocol: Lessons for Formulating Policies for Global Warming. *Policy Studies Journal* 19:152–161.

Morrow, James D. 1994. Modeling the Forms of International Cooperation: Distribution Versus Information. *International Organization* 48:387–423.

Mortimore, Michael. 1993. The Sahel. In *International Environmental Negotiation*, ed. G. Sjöstedt. Newbury Park, CA: Sage Publications.

Most, Benjamin A. and Harvey Starr. 1989. *Inquiry, Logic, and International Politics*. Columbia, SC: University of South Carolina Press.

Mouawad, Jad. 2006. No U.S. Rules, But Some Firms Reduce Emissions. *New York Times* (May 29). Web address: http://www.iht.com/articles/2006/05/29/business/carbon.php.

Mukela, John. 1992a. Africa Presses for Action. *Crosscurrents* (March 9–11): 16.

———. 1992b. Forests: In Search of Principles. *Development Forum* (May–June): 11.

Multilateral Fund for the Implementation of the Montreal Protocol Secretariat. 2005. Post Meeting Summary of Decisions of the 45th Meeting of the Executive Committee of the Multilateral Fund for the Implementation of the Montreal Protocol. http://www.multilateralfund.org/files/PMS45.pdf.

Myers, Norman. 1984. *The Primary Source: Tropical Forests and Our Future*. New York: W.W. Norton.

———. 1990. Tropical Forests. In *Global Warming: The Greenpeace Report*, ed. J. Leggett. Oxford, UK: Oxford University Press.

Nations Have Sovereignty over Forests. 1992. *Jornal do Brasil* (June 13): 6.

Naughton, Philippe. 2005. Asia-Pacific Climate Pact Takes UK by Surprise. *Times Online* (July 28). http://www.timesonline.co.uk/article/0, 2–1711816, 00.html.

Netter, Thomas W. 1987. U.N. Parley Agrees to Protect Ozone. *New York Times* (May 1): A1.

Nicholson, Michael. 1996. The Continued Significance of Positivism. In *International Theory: Positivism and Beyond*, ed. K.B. Steve Smith and Marysia Zalewski. Cambridge, UK: Cambridge University Press.

———. 2002. *International Relations: A Concise Introduction, 2nd ed.* New York: New York University Press.

Niou, Emerson M.S. and Peter C. Ordeshook. 1991. Realism versus Neoliberalism: A Formulation. *American Journal of Political Science* 35:481–511.

Nixon, Richard. 1992. Yeltsin Clearly Deserves Help. *Earth Summit Times* (June 13): 15.

North, Richard. 1989. Appeal for Fund to Help Third World Cut CFCs. *The Independent* (March 8): 6.

Nuisance Case: States and NGOs Appeal Dismissal (December 15, 2005). Climate Justice Programme, 2005/12/22. http://www.climatelaw.org/media/U.S.%20 nuisance%20appeal.

Oberthür, Sebastian. 1997. Montreal Protocol: 10 Years After. *Environmental Policy and Law* 27:432–441.

———. 1999. The EU as an International Actor: The Protection of the Ozone Layer. *Journal of Common Market Studies* 37:641–659.

———. 2000. Ozone Layer Protection at the Turn of the Century: The Eleventh Meeting of the Parties. *Environmental Policy and Law* 30:34–41.

Oberthür, Sebastian and Hermann Ott. 1999. *The Kyoto Protocol: International Climate Policy for the 21st Century*. New York: Springer.

Odell, John S. 2001. Case Study Methods in International Political Economy. *International Studies Perspectives* 2:161–176.

Olson, Mancur Jr. 1965. *The Logic of Collective Action*. Cambridge, MA: Harvard University Press.

Ostrom, Elinor. 1990. *Governing the Commons: The Evolution of Institutions for Collective Action*. Cambridge, MA: Harvard University Press.

———. 1996. Institutional Rational Choice: An Assessment of the IAD Framework. Paper presented at the American Political Science Association Annual Conference, San Francisco, CA, August 29–September 1, 1996.

Oye, Kenneth A. 1986. Explaining Cooperation Under Anarchy: Hypotheses and Strategies. In *Cooperation Under Anarchy*, ed. K.A. Oye. Princeton, NJ: Princeton University Press.

Oye, Kenneth A. and James H. Maxwell. 1994. Self-Interest and Environmental Management. *Journal of Theoretical Politics* 6:593–624.

Panjabi, Ranee K.L. 1997. *The Earth Summit at Rio: Politics, Economics, and the Environment*. Boston, MA: Northeastern University Press.

Park, Gary. 2006. Kyoto has Canada in Knots: New Regime Pulls Back from Kyoto Spending; Confusion on Made-In Canada Policy. *Petroleum News* (May 14). http://www.petroleumnews.com/pntruncate/691933951.shtml.

Parson, Edward A. 1993. Protecting the Ozone Layer. In *Institutions for the Earth: Sources of Effective International Environmental Protection*, ed. Peter M. Haas, Robert O. Keohane, and Marc A. Levy. Cambridge, MA: MIT Press.

Parson, Edward A. 2003. *Protecting the Ozone Layer: Science and Strategy*. Oxford University Press: Oxford.

Paterson, Mathew. 1992. Global Warming. In *The Environment in International Relations*, ed. C. Thomas. London: Royal Institute of International Affairs.
———. 1994. Review Article on Institutions For the Earth. *International Affairs* 70:153.
———. 1996. IR Theory. In *The Environment and International Relations*, ed. John Vogler and Mark F. Imber, 59–76. London: Routledge.
———. 2000. Car Culture and Global Environmental Politics. *Review of International Studies* 26:253–270.
———. 2001. Climate Policy as Accumulation Strategy: The Failure of COP-6 and Emerging Trends in Climate Politics. *Global Environmental Politics* 1:10–17.
Paterson, Matthew and Michael Grubb. 1992. The International Politics of Climate Change. *International Affairs* 68:293–310.
Pearce, Fred. 1989. Kill or Cure? Remedies for the Rainforest. *New Scientist* (September 16): 40–43.
———. 1991. Road to Rio Strewn with Cant. *New Scientist* (October 19): 12.
———. 1997. Forum: Captain Eco Rides Again. *New Scientist* (October 4): 46.
———. 2005. European Trading in Carbon-Emission Permits Begins. *New Scientist* (January 6). http://www.newscientist.com/article.ns?id=dn6846.
Peterson, M.J. 1998. Organizing for Effective Environmental Cooperation. *Global Governance* 4:415–438.
———. 1999. Many Hands Make Green Work. *International Studies Review* 1:83–93.
Pettigrew Calls on U.S. to Respond to Canada's Proposals in Softwood Dispute. Canadian Press, February 4, 2002. http://ca.news.yahoo.com/020203/6/icmx.html.
Pianin, Eric. 2001. U.S. Aims to Pull Out of Warming Treaty; "No Interest" in Implementing Kyoto Pact, Whitman Says. *Washington Post* (March 28): A01.
Pitt, David E. 1992a. Forest . . . Finance . . . Frustration. *Earth Summit Times* (June 12): 1, 16.
———. 1992b. U.S. Pulls All Stops for "Forest Principles." *Earth Summit Times* (June 10): 1, 16.
Pollution and the Poor, Why "Clean Development" at any Price is a Curse on the Third World. 1992. *The Economist* (May 30): 9–10.
Pope, Justin. 2004. Greenspan Urges Education and Training to Address Global Job Loss Worries. *The Detroit News* (March 13). http://www.detnews.com/2004/business/0403/14/business-90451.htm.
Porter, Gareth and Inji Islam. 1992. The Road from Rio: An Agenda for U.S. Follow-up to the Earth Summit. Washington, DC: Environmental and Energy Study Institute.
Porter, Gareth and Janet Welsh Brown. 1991. *Global Environmental Politics*. Boulder, CO: Westview Press.
———. 1996. *Global Environmental Politics*, 2nd ed. Boulder, CO: Westview Press.
Porter, Gareth, Janet Welsh Brown, and Pamela S. Chasek. 2000. *Global Environmental Politics*. Boulder, CO: Westview Press.
"Possible Main Elements of an Instrument (Convention, Agreement, Protocol, Charter, etc.) for the Conservation and Development of the World's Forests" (FAO draft), Rome, September 18, 1990.

Prager, Herman. 1992. The Ecosystem and Advanced Industrial Society: Is a Change Occurring in the American Strategy Towards UNCED? Paper presented at the International studies Association Annual Convention, Atlanta, GA, March 31–April 4, 1992.

PrepCom II: What Was Accomplished? *Earth Summit Update* (July): 1.

Press Release. 1987. Council Decision on the Protection of the Ozone Layer is "Eye-Wash." European Environmental Bureau.

Prowse, Michael and David Lascelles. 1992. Financing a Green Future in a Planet without Borders. *The Financial Times* (February 14): 5.

Pyle, J.A., S. Solomon, D. Wuebbles, and S. Zvenigorodsky. 1992. Ozone Depletion and Chlorine Loading Potentials. In *Scientific Assessment of Ozone Depletion: 1991*. World Meteorological Organization Global Ozone Research and Monitoring Project—Report No. 25. Geneva: World Meteorological Organization. http://www/ciesin.org/docs/011-551/011-551.html.

Raghavan, Chakravarthi. 1992. Stumbles on Finance, Forests, Air, and Deserts. *SUNS at the Earth Summit* (June 19): 1–2.

Raiffa, Howard. 1982. *The Art and Science of Negotiation*. Cambridge, MA: Belknap Press of Harvard University Press.

Ramakrishna, Kilaparti. 1992. North-South Issues, the Common Heritage of Mankind and Global Environmental Change. In *Global Environmental Change and International Relations*, ed. Ian H. Rowlands and Malory Greene. Basingstoke, UK: Macmillan.

RAND ES&PC. 2006. Alternative Fluorocarbons Environmental Acceptability Study Research and Assessment Program. http://www.afeas.org/about.html.

Rapoport, Anatol and Melvin Guyer. 1966. A Taxonomy of 2 × 2 Games. *General Systems* 11:203–214.

Ravilious, Kate. 2005. Global Warming: Death in the Deep-Freeze. *The Independent* (September 28): 44–45.

Rawls, John. 1971. *A Theory of Justice*. Cambridge, MA: Belknap Press of Harvard University Press.

Redclift, Michael R. 1993. *Sustainable Development: Exploring the Contradictions*. London: Methuen.

Regional Greenhouse Gas Initiative: An Initiative of the Northeast and Mid-Atlantic States. "Participating States." http://www.rggi.org/states.htm.

Reuters. 2005. Antarctic Ozone Hole Gets Bigger. *The Australian* (August 25): 8.

Revkin, Andrew C. 2005. In a Melting Trend, Less Arctic Ice to Go Around. *New York Times* (September 29): 1.

———. Climate Expert Says NASA Tried to Silence Him. *New York Times* (January 29): 1.

Roan, Sharon. 1989. *Ozone Crisis: The 15-year Evolution of a Sudden Global Emergency*. New York: Wiley.

Rock Ethics Institute. 2005. Ethical Dimensions of Climate Change. Pennsylvania State University. http://rockethics.psu.edu/climate/index.htm.

Root of Evil at Rio. 1992. *The Economist* (June 13): 12–13.

Rowlands, Ian H. 1995. *The Politics of Global Atmospheric Change*. New York: St. Martin's Press.

———. 2001. Classical Theories of International Relations. In *International Relations and Global Climate Change*, ed. Urs Luterbacher and Detlef F. Sprinz. Cambridge, MA: MIT Press.

Rowlands, Ian and Malory Greene. 1992. *Global Environmental Change and International Relations*. Basingstoke, UK: Macmillan.

Rubin, Jeffrey Z. 1993. Third Party Roles: Mediation in International Environmental Disputes. In *International Environmental Negotiation*, ed. G. Sjöstedt. Newbury Park, CA: Sage Publications.

Runnalls, David. 1992a. Bush Rio Talk to Stress Forest. *Earth Summit Times* (June 1): 1.

———. 1992b. Summit Recap: No Cash, More G-77. *Earth Summit Times* (June 14): 3.

Ryan, John C. 2002. On Climate, States Lead. *Christian Science Monitor* (July 25): 8.

Sachs, Wolfgang. 1991. Environment and Development: The Story of a Dangerous Liaison. *The Ecologist* (November/December): 252–257.

Sand, Peter H. 1990. *Lessons Learned in Global Environmental Governance*. Washington, DC: World Resources Institute.

———. 1992. *The Effectiveness of International Environmental Agreements: A Survey of Existing Legal Instruments*. Cambridge, UK: Grotius Publishers Limited.

———. 1993. International Environmental Law After Rio. *European Journal of International Law* 4:390–417.

Sandler, Todd. 1997. *Global Challenges: An Approach to Environmental, Political, and Economic Problems*. Cambridge, UK: Cambridge University Press.

Saunders, Phillip M. 1996. Development Assistance Issues Related to a Convention on Forests. In Canadian Council on International Law. *Global Forests and International Environmental Law*. London: Kluwer Law International.

Schelling, Thomas C. 1956. An Essay on Bargaining. *The American Economic Review* 46:281–306.

———.1978. *Micromotives and Macrobehavior*. New York: W.W. Norton.

Schneider, Keith. 1992. The Earth Summit; White House Snubs US Envoy's Plea to Sign Rio Treaty. *New York Times* (June 5): A1.

Schoppa, Leonard J. 1999. The Social Context in Coercive International Bargaining. *International Organization* 53:307–342.

Schröder, Heike. 2001. *Negotiating the Kyoto Protocol: An Analysis of Negotiation Dynamics in International Relations*. Münster, Germany: Lit Verlag.

Schulte, Bret. 2006a. Saying It In Cinema. *U.S. News & World Report* (June 5): 38–39.

———. 2006b. Special Report: Temperature Rising. *U.S. News & World Report* (June 5): 36–43.

Schweller, Randall L. and David Priess. 1997. A Tale of Two Realisms: Expanding the Institutions Debate. *Mershon International Studies Review* 41:1–32.

Sebenius, James K. 1983. Negotiation Arithmetic: Adding and Subtracting Issues and Parties. *International Organization* 37:281–316.

———. 1991a. Designing Negotiations Toward a New Regime: The Case of Global Warming. *International Security* 15:110–148.

———. 1991b. Negotiation Analysis. In *International Negotiation: Analysis, Approaches, Issues*, ed. V.A. Kremenyuk and G. Faure. San Francisco: Jossey-Bass Publishers.

———. 1992a. Challenging Conventional Explanations of International Cooperation: Negotiation Analysis and the Case of Epistemic Communities. *International Organization* 46:323–365.

———. 1992b. Negotiation Analysis: A Characterization and Review. *Management Science* 38:18–38.

———. 1993. The Law of the Sea Conference: Lessons for Negotiations to Control Global Warming. In *International Environmental Negotiation*, ed. G. Sjöstedt. Newbury Park, CA: Sage Publications.

———. 1995. Overcoming Obstacles to a Successful Climate Convention. In *Shaping National Responses to Climate Change*, ed. H. Lee. Washington, DC: Island Press.

Seelye, Katherine Q. 2002. President Distances Himself from Global Warming Report. *New York Times* (June 5): 23.

Sell, Susan. 1996. North-South Environmental Bargaining: Ozone, Climate Change, and Biodiversity. *Global Governance* 2:97–118.

Senior UN Officials, Pension Fund Heads, CEOs, Wall Street Leaders to Discuss Climate Risks, Opportunities at Summit, May 10, 2005. UN Press Release Note No. 5938 (May 6).

Shea, Cynthia Pollock. 1988. *Protecting Life on Earth: Steps to Save the Ozone Layer*. Washington, DC: Worldwatch Institute.

Shimberg, Steven J. 1991. Stratospheric Ozone and Climate Protection: Domestic Legislation and the International Process. *Environmental Law* 21:2175–2216.

Shin, Paul H. B. 2005. White House's Icy Reception to Climate Threats not Stopping Some Cities. *New York Daily News* (December 11): 26.

Simmons, Marlise. 1992. North-South Chasm is Threatening Search for Environmental Solutions. *New York Times* (March 17): A5.

Singer, J. David. 1961. The Level-of-Analysis Problem in International Relations. *World Politics* 14:77–92.

Singer, S. Fred. 1992. Earth Summit Will Shackle the Planet, Not Save It. *Wall Street Journal*, (February 19): A14.

Sjöstedt, Gunnar. 1993. *International Environmental Negotiation*. Newbury Park, CA: Sage Publications.

Skolnikoff, Eugene. B. 1990. The Policy Gridlock on Global Warming. *Foreign Policy* (Summer): 77–94.

Smith, Brigitte. 1998. Ethics of Du Pont's CFC Strategy 1975–1995. *Journal of Business Ethics* 17:557–568.

Smith, H.A. 1992. An Environmental Meta-Regime. Paper presented at the International Studies Association Annual Convention, Atlanta, GA, March 31–April 4, 1992.

Snidal, Duncan. 1985a. Coordination vs. Prisoner's Dilemma: Implications for International Cooperation and Regimes. *American Political Science Review* 79:923–942.

———. 1985b. The Limits of Hegemonic Stability Theory. *International Organization* 39:579–614.

Snidal, Duncan. 1990. IGOs, Regimes, and Cooperation: Challenges for International Relations Theory. In Margaret P. Karns and Karen A. Mingst. *The United States and Multilateral Institutions: Patterns of Changing Instrumentality and Influence*. Boston: Unwin Hyman.

Society of American Foresters. 1999. Task Force Report on Forest Management Certification Programs, 1999 Report. http://www.safnet.org/policyandpress/fmcp1999.doc.

Solomon, Susan and Daniel Albritton. 1992. Time-Dependent Ozone Depletion Potentials for Short- and Long-Term Forecasts. *Nature* (May 7): 33–37.

Solow, Robert. 1992. An Almost Practical Step Toward Sustainability. Lecture presented for the Fortieth Anniversary of Resources for the Future, October 8, 1992, Washington, DC.

Soroos, Marvin S. 1998. The Assault on Tropical Rain Forests. *Mershon International Studies Review* 42:317–321.

———. 2001. Global Climate Change and the Futility of the Kyoto Process. *Global Environmental Politics* 1:1–9.

Spector, Bertram I. 1992. *Post Negotiation- Is the Implementation of Future Negotiated Environmental Agreements Threatened?: A Pilot Study*. Laxenburg, Austria: International Institute for Applied Systems Analysis.

Sprinz, Detlef. 1991. International Environmental Security: Individual vs. Collective Rationality. Paper presented at the International Studies Association Annual Convention, Vancouver, British Columbia, March 20–23, 1991.

———. 1994. Editorial Overview: Strategies of Inquiry into International Environment Policy. *International Studies Notes* 19:32–57.

———. 1999. Research on the Effectiveness of International Regimes: A Review of the State of the Art. Paper presented at the International Studies Association Annual Conference, February 16–20, 1999, Washington, DC.

Sprinz, Detlef and Tapani Vaahtoranta. 1994. The Interest-Based Explanation of International Environmental Policy. *International Organization* 48:77–106.

Sprinz, Detlef F. and Martin Weiβ. 2001. Domestic Politics and Global Climate Policy. In *International Relations and Global Climate Change*, ed. Urs Luterbacher and Detlef F. Sprinz, 67–94. Cambridge, MA: MIT Press.

Starkman, Dean. 2005. A New Worry for Insurers; Firms Looking at Whether Climate Change Could Affect Their Bottom Lines. *The Washington Post* (October 5): D1.

Starnes, Richard. 1997. Scientists say UN Bureaucrats Rewrote Findings. *The Ottawa Citizen* (August 28): A1.

Statement of Cecil E. Roberts, President of the United Mine Workers of America before the United States Senate Committee on Energy and Natural resources on the Economic Impacts of the Kyoto Protocol, March 25, 1999. http://www.umwa.org/legaction/globalwarming/statement.shtml.

Stein, Arthur A. 1983. Coordination and Collaboration: Regimes in an Anarchic World. In *International Regimes*, ed. Stephen D. Krasner. Ithaca, NY: Cornell University Press.

Stevis, Dimitris and Valerie J. Assetto, eds. 2001. *The International Political Economy of the Environment: Critical Perspectives. International Political Economy Yearbook, Vol.* 12. Boulder, CO: Lynne Rienner Publishers.

Steyer, Robert. 1997a. Growers Brace for Jolt from Fumigant Ban. St. Louis *Post Dispatch* (May 5): 12.

———. 1997b. Strawberry Fields Forgotten? *Chicago Sun-Times* (May 11): 66.

Stiglitz, Joseph E. 2000. *Economics of the Public Sector*. New York: W.W. Norton.

Stolarski, Richard S. and Ralph J. Cicerone. 1974. Stratospheric Chlorine: A Possible Sink for Ozone. *Canadian Journal of Chemistry* 52:1610–1615.

Stone, Christopher D. 1992. Beyond Rio: "Insuring" Global Warming. *American Journal of International Law* 86:445–488.

Strange, Susan. 1983. Cave! Hic Dragones: A Critique of Regime Analysis. In *International Regimes*, ed. Stephen Krasner. Ithaca, NY: Cornell University Press.

Sullivan, Francis. 1993. Forest Principles. In *The "Earth Summit" Agreements: A Guide and Assessment*, ed. Michael Grubb, Matthias Koch, Koy Thomson, Abby Munson, and Francis Sullivan. London: Earthscan Publications Ltd.

Suplee, Curt. 1998. Gas Buildup Could Lead to Arctic "Ozone Hole". *Washington Post* (April 9): A14.

Susskind, Lawrence. 1994. *Environmental Diplomacy: Negotiating More Effective Global Agreements*. New York: Oxford University Press.

Sussman, Glen. 2004. The USA and Global Environmental Policy: Domestic Constraints on Effective Leadership. *International Political Science Review* 25:349–369.

Svensson, Ulf. 1993. The Convention on Biodiversity: A New Approach. In *International Environmental Negotiations*, ed. G. Sjöstedt. Newbury Park, CA: Sage Publications.

Szell, Patrick. 1993. Negotiations on the Ozone Layer. In *International Environmental Negotiation*, ed. G. Sjöstedt. Newbury Park, CA: Sage Publications.

Taib, Fauziah Mohd. 1997. *Malaysia and UNCED: An Analysis of a Diplomatic Process: 1989–1992*. London: Kluwer International Press.

Taylor, Michael. 1987. *The Possibility of Cooperation*. Cambridge, UK: Cambridge University Press.

Technology Brief—Du Pont Co.: Ozone-Friendly Refrigerants Are Set for Commercial Use. *Wall Street Journal* (September 23): B6.

The 1992 Rio Earth Summit: Objectives and Achievements. *Earth Summit Times* (August 27): 9.

The ASEAN Ministers for the Environment. 1990. The Kuala Lumpur Accord on Environment and Development. Kuala Lumpur: Association of Southeast Asian Nations.

The Environment, International Relations, and U.S. Foreign Policy. Washington, DC: Georgetown University Press.

The Montreal Protocol: A Briefing Book. Rosslyn, VA: Alliance for Responsible CFC Policy, 1987.

"The Most Important Issue that We Face." 2005. *The Independent* (April 18): 1.

Thirteen Pension Leaders Call on SEC Chairman to Require Global Warming Risks in Corporate Disclosures. 2004. Ceres Press Release (April 15).

Thomas, Caroline.1992. *The Environment in International Relations*. London: Royal Institute of International Affairs.

Tolba, Mostafa Kamal and Iwona Rummel-Bulska. 1998. *Global Environmental Diplomacy: Negotiating Environmental Agreements for the World, 1973–1992*. Cambridge, MA: MIT Press.

Too Much Too Fast. 1992. Newsweek (June 1): 34.
Trail Smelter Case, United States of America versus Canada 1938 and 1941. In UNEP/UNDP/Dutch Joint Project on Environmental Law and Institutions in Africa. *Compendium of Judicial Decisions on Matters Related to Environment. International Decisions, Vol. 1*. December 1998, 1.
Trevison, Catherine. 2005. Suit Over Emission Threat Hovers in Legal Gray Zone. *The Oregonian* (October 5): C01.
Tversky, Amos, and Daniel Kahneman. 1986. Rational Choice and the Framing of Decisions. *The Journal of Business* 59:S251.
U.S. is Assailed at Geneva Talks for Backing Out of Ozone Plan. 1990. *New York Times* (May 10): A1.
U.S. May Veto Key Parts of Agenda 21. 1992. *Earth Summit Update* (March): 1, 3.
U.S. Principles Downplay Primary Forests. 1991. *Earth Summit Update* (July): 1.
U.S. Rejects Targets in Forest Negotiations. 1991. *Earth Summit Update* (September): 2.
U.S. Singled Out as Eco Bad Guy. 1992. *Jornal do Brasil* (June 5): 1, 6.
U.S. to Join Fund to Help Curb Ozone Depletion. 1990. *Los Angeles Times* (June 16): A27.
U.S. Will Oppose Key Agenda 21 Options at PrepCom 3. 1991. *Earth Summit Update* (August): 1–2.
Un Susurro en el Bosque. 1992. *Crosscurrents* (June 5): 22.
Underdal, Arild. 1994. Leadership Theory: Rediscovering the Arts of Management. In *International Multilateral Negotiation: Approaches to the Management of Complexity*, ed. W.I. Zartman. San Francisco: Jossey-Bass Publishers.
UNDP. 2005a. Managing Chemicals Sustaining Livelihoods: UNDP and Management of Persistent Organic Pollutants, Ozone-Depleting Substances and Other Chemicals. Phasing Out Ozone-Depleting Substances Under the Montreal Protocol. http://www.undp.org/montrealprotocol/Publications/Chemicals_booklet.pdf.
———. 2005b. Phasing Out Ozone-Depleting Substances under the Montreal Protocol. http://www.undp.org/gef/undp-gef_publications/publications/chemicalspops_april2005.pdf.
UNEP. 1983. Draft Annex Concerning Measures to Control, Limit and Reduce the Use and Emissions of Fully Halogenated Chlorofluorocarbons (CFCs) for the Protection of the Ozone Layer (UNEP/WG.94/4). *Ad Hoc* Working Group of Legal and Technical Expert for the Elaboration of a Global Framework Convention for the Protection of the Ozone Layer, Third Session, Geneva, October 17–21. Nairobi: United Nations Environment Programme.
———. 1985. Vienna Convention for the Protection of the Ozone Layer. Nairobi: United Nations Environment Programme Ozone Secretariat. http://ozone.unep.org/pdfs/viennaconvention2002.pdf.
———. 1989a. Report of the Parties to the Montreal Protocol on the Work of their First Meeting (UNEP/OzL.Pro.1/5). Nairoba: United Nations Environment Programme Ozone Secretariat. http://ozone.unep.org/Meeting_Documents/mop/01mop/1mop-5e.shtm.
———. 1989b. Synthesis Report (UNEP/OzL.Pro.WG.II(1)/4). Nairobi: United Nations Environment Programme Ozone Secretariat. http://ozone.unep.org/Meeting_Documents/oewg/2oewg/2oewg1-4.e.doc.

———. 1990. Report of the Third Meeting of the Bureau of the Montreal Protocol (UNEP/OzL.Pro.Bur.3/2). Nairobi: United Nations Environment Programme Ozone Secretariat.

———. 1992. Report of the Fourth Meeting of Parties to the Montreal Protocol (UNEP/OzL.Pro.4/15). Nairobi: United Nations Environment Programme Ozone Secretariat. http://ozone.unep.org/Meeting_Documents/mop/04mop/4mop-15.e.doc.

———. 1994. 1994 Report of the Methyl Bromide Technical Options Committee for the 1995 Assessment of the Montreal Protocol on substances that Deplete the Ozone Layer. Nairobi, Kenya: United Nations Environment Programme Ozone Secretariat. http://ozone.unep.org/teap/Reports/MBTOC/MBTOC94.pdf.

———. 1995a. Report of the Seventh Meeting of the Parties to the Montreal Protocol (UNEP/OzL.Pro.7/12). Nairobi: United Nations Environment Programme Ozone Secretariat. http://ozone.unep.org/Meeting_Documents/mop/07mop/7mop-12.e.doc.

———. 1995b. Technology and Economic Assessment Panel: Report to the Parties. Nairobi: United Nations Environment Programme Ozone Secretariat.

———. 1997a. Report of the Ninth Meeting of the Parties to the Montreal Protocol on Substances that Deplete the Ozone Layer (UNEP/OzL.Pro.9/12). Nairobi: United Nations Environment Programme Ozone Secretariat. http://ozone.unep.org/Meeting_Documents/mop/09mop/9mop-12.e.doc.

———. 1997b. Report of the Technical and Economic Assessment Panel. Vol. 1: April. Nairobi: United Nations Environment Programme Ozone Secretariat. http://ozone.unep.org/teap/Reports/index.asp.

———. 1997c. Corrigendum to April 1997 TEAP Report. http://www.unep.org/ozone/teap/Reports/TEAP_Reports/Crigndm.pdf.

———. 1998. Report of the Tenth Meeting of the Parties to the Montreal Protocol (UNEP/OzL.Pro.10/9). Nairobi: United Nations Environment Programme Ozone Secretariat. http://ozone.unep.org/Meeting_Documents/mop/10mop/10mop-9.e.doc.

———. 1999a. GEO-2000 (Global Environment Outlook 2000). Nairobi: United Nations Environment Programme. http://www.unep.org/Geo2000/english/0231.htm.

———. 1999b. Report of the Eleventh Meeting of the Parties to the Montreal Protocol (UNEP/OzL.Pro.11/10). Nairobi: United Nations Environment Programme Ozone Secretariat. http://ozone.unep.org/Meeting_Documents/mop/11mop/11mop-10.e.doc.

———. 1999c. Synthesis of the 1998 Scientific, Environmental Effects, and Technology and Economic Assessments (UNEP/OzL.Pro/WG.1/19/3). Nairobi: United Nations Environment Programme Ozone Secretariat. http://ozone.unep.org/Meeting_Documents/oewg/19oewg/19oewg-3.e.htm.

———. 1999d. Technology and Economic Assessment Panel Supplementary Report to "Assessment of the Funding Requirement for the Replenishment of the Multilateral Fund for the Period of 2000–2002." Nairobi: United Nations Environment Programme Ozone Secretariat. http://ozone.unep.org/teap/Reports/TEAP_Reports/FUND99S.pdf.

UNEP. 2000a. *Action on Ozone: 2000 Edition*. Nairobi: United Nations Environment Programme Ozone Secretariat. http://ozone.unep.org/pdfs/ozone-action-en.pdf.

———. 2000b. The Implications of Becoming or Not Becoming a Party to the Vienna Convention for the Protection of the Ozone Layer and the Montreal Protocol on Substances that Deplete the Ozone Layer, including its Amendments. Nairobi: United Nations Environment Programme Ozone Secretariat. http://www.unep.org/ozone/12mop-inf5.shtml.

———. 2000c. The Montreal Protocol on Substances that Deplete the Ozone Layer, as Adjusted and/or Amended in London 1990, Copenhagen 1992, Vienna 1995, Montreal 1997, and Beijing 1999. Nairobi: United Nations Environment Programme Ozone Secretariat. http://ozone.unep.org/pdfs/Montreal-Protocol2000.pdf.

———. 2003a. Backgrounder: Basic Facts and Data on the Science and Politics of Ozone Depletion. Nairobi: United Nations Environment Programme. http://hq.unep.org/ozone/pdf/Press-Backgrounder.pdf.

———. 2003b. Report of the Technology and Economic Assessment Panel May 2003 HCFC Task Force Report. Nairobi: United Nations Environment Programme Ozone Secretariat. http://ozone.unep.org/teap/Reports/Other_Task_Force/HCFC03R1.pdf.

———. 2004a. Chemicals Phase Out Schedule. Nairobi: United Nations Environment Programme Ozone Secretariat. http://ozone.unep.org/Public_Information/4Aiv_PublicInfo_Facts_chemicals.asp.

———. 2004b. Informal Consultation on Methyl Bromide: Buenos Aires, March 4–5, 2004. Nairobi: United Nations Environment Programme Ozone Secretariat. http://www.unep/org/ozone/meeting_documents/mop/Ex_mop/1ex_mop-inf1.e.pdf.

———. 2004c. Kyoto Protocol to Enter into Force February 16, 2005. Nairobi: United Nations Environment Programme. http://www.unep.org/Documents.Multilingual/Default.Print.asp?DocumentID=412&ArticleID=4669&l=en.

———. 2004d. Report of the First Extraordinary Meeting of the Parties to the Montreal Protocol on Substances that Deplete the Ozone Layer (UNEP/OzL.Pro.ExMP/1/3). Nairobi: United Nations Environment Programme Ozone Secretariat. http://ozone.unep.org/Meeting_Documents/mop/Ex_mop/1ex_mop-3.e.doc.

———. 2004e. Report of the Technology and Economic Assessment Panel. Critical Use Nominations for Methyl Bromide Final Report. Nairobi: United Nations Environment Programme Ozone Secretariat. www.unep.org/ozone/teap/Reports/MBTOC/MBCUN-october2004.pdf.

———. 2004f. Summary Control Measures. Nairobi: United Nations Environment Programme Ozone Secretariat, http://ozone.unep.org/Treaties_and_Ratification/2Biii_1summary_controls_measures.asp.

———. 2005a. New Report on How to Save the Ozone Layer while Combating Climate Change. Nairobi: United Nations Environment Programme. http://www.unep.org/Documents.Multilingual/Default.asp? DocumentID=430&ArticleID=4768&l=en.

———. 2005b. Trade Names of Chemicals Containing Ozone Depleting Substances and their Alternatives. The OzonAction Programme Library. Nairobi: United

Nations Environment Programme. http://www.uneptie.org/ozonaction/library/tradenames/trade_chem.asp?id=87.

———. 2006a. GEO: Global Environment Outlook. 2006. *GEO Year Book 2006*. Nairobi: United Nations Environment Programme. http://www.unep.org/geo/yearbook/yb2006/009.asp.

———. 2006b. Summary of Issues for Discussion by the Open-ended Working Group of the Parties to the Montreal Protocol: Note by the Secretariat (UNEP/OzL.Pro.WG.1/26/2). Nairobi: United Nations Environment Programme. http://ozone.unep.org/Meeting_Documents/oewg/26oewg/OEWG-26-2E.pdf.

———. 2006c. Table: Status of Ratification. Nairobi: United Nations Environment Programme Ozone Secretariat. http://ozone.unep.org/Treaties_and_Ratification/2C_ratificationTable.asp.

United Nations. 1991. Ad Hoc Subgroup, Forests, Draft Synoptic List (A/CONF.151/PrepCom/WG.I/Misc.3).

———. 1992. Report of the United Nations Conference on Environment and Development: Annex III: Non-Legally Binding Authoritative Statement of Principles for a Global Consensus on the Management, Conservation and Sustainable Development of All Types of Forests (A/CONF.151/26 (Vol. III). New York: United Nations.

———. 1992. The Rio Declaration on Environment and Development. New York: United Nations.

———. 1994. United Nations Convention to Combat Desertification in Countries Experiencing Serious Drought and/or Desertification, Particularly in Africa. http://www.unccd.int/convention/text/convention.php.

United Nations Environment Programme, World Resources Institute, World Bank, and United Nations Development Programme. 2000. *World Resources 2000–2001: People and Ecosystems: the Fraying Web of Life*. Washington, DC: World Resources Institute.

United Nations Food and Agriculture Organization (UNFAO). 1990. DRAFT: Possible Main Elements of an Instrument (Convention, Agreement, Protocol, Charter, etc.) for the Conservation and Development of the World's Forests. Rome: UNFAO.

———. 2005. *Global Forest Resources Assessment 2005*. http://www.fao.org/forestry/site/fra2005/en.

United Nations Framework Convention on Climate Change. 1992. http://unfccc.int/essentialbackground/convention/background/items/1349.php.

———. 1997a. Kyoto Protocol to the United Nations Framework Convention on Climate Change (FCCC/CP/1997/L.7/Add.1). http://unfccc.int/cop3/l07a01.htm. New York: United Nations.

———. 1997b. Kyoto Protocol to the United Nations Framework Convention on Climate Change: Final Draft by the Chairman of the Committee of the Whole (FCCC/CP/1997/CRP.6). New York: United Nations.

———. 2005a. Current Evidence of Climate Change. http://unfccc.int/essential_background/feeling_the_heat/items/2904.php.

———.2005b. Decision 27/CMP.1: Procedures and Mechanisms Relating to Compliance under the Kyoto Protocol. In Report of the Conference of the Parties Serving as the Meeting of the Parties to the Kyoto Protocol on its First Session, held at Montreal from November 28 to December 10, 2005, Addendum: Action Taken by the Conference of the Parties Serving as the Meeting of the Parties to the Kyoto Protocol at its First Session (FCCC/KP/CMP/2005/8/Add.3). http://unfccc.int/resource/docs/2005/cmp1/eng/08a03.pdf#page=92.

United Nations Framework Convention on Climate Change. 2005c. The Greenhouse Effect and the Carbon Cycle. http://unfccc.int/ essential_background/feeling_the_heat/items/2903txt.php.

———. 2006. Kyoto Protocol Status of Ratification. April 18, 2006. http://unfcc.int/files/essential_background/kyoto_protocol/application/pdf.kpstats.pdf.

United Nations General Assembly. 1990. Resolution Adopted by the General Assembly [on the report of the Second Committee (A/44/746/Add.7)]: New York: United Nations.

United States. 1990a. Unpublished Draft. An International Convention on the World's Forests. July 5, 1990.

———. 1990b. US Talking Points on the Forestry Convention. Unpublished Manuscript.

United States. 1991. Statement on Forests to PrepCom 2. Geneva. Unpublished Manuscript.

United States Census Bureau. 2005. Statistical Abstract of the United States 2004–2005. Table 851. Timber Products–Production, Foreign Trade, and Consumption by Type of Product: 1990 to 2004. http://www.census.gov/prod/2005pubs/06states/natresor.pdf.

United States Congress. 1986. *Congressional Record: Proceedings and Debates of the Congress*. October 8, 1986, S15679. Washington, DC.

———. 1987. *Congressional Record: Proceedings and Debates of the Senate*. June 5, 1987, S7712. Washington, DC.

———. 2005. Office of Technology Assessment Oceans and Environment Program. 1988. An Analysis of the Montreal Protocol on Substances that Deplete the Ozone Layer. Washington, DC.

———. 2005. Energy Policy Act of 2005, January 4, 2005. http://frwebgate.access.gpo.gov/cgi-bin/getdoc.cgi?dbname=109_cong_bills &docid=f:h6enr.txt.pdf.

United States Congress etc. 2005. Congressional Budget Office. 2005. Cost Estimate for the Bill Conference Agreement, H.R. 6, Energy Policy Act of 2005 (July 27). http://www.cbo.gov/showdoc.cfm?index=6581&sequence=0&from=6.

US Department of Energy. 1997. *Mitigating Greenhouse Gas Emissions: Voluntary Reporting*. Energy Information Administration, Office of Integrated Analysis and Forecasting. Washington, DC, October.

US Department of Energy, Energy Information Administration. 1998. What Does the Kyoto Protocol Mean to the U.S. Energy Markets and the U.S. Economy? Report #: SR/OIAF/98-03, October 1998. http://www.eia.doe.gov/oiaf/ kyoto/ kyototext.html.

USDA Forest Service. 2004. National Report on Sustainable Forests—2003. Doc. no. FS-766. Washington, DC.

United States Environmental Protection Agency, Office of Air and Radiation and World Resources Institute. 1995. *Protection of the Ozone Layer*. EPA Environmental Ozone Indicators, EPA 230-N.95-002, EPA June 1995. Washington, DC.

United States Environmental Protection Agency (USEPA). 1987a. An Assessment of the Risks of Stratospheric Modification. Washington, DC.

———. 1987b. Regulatory Impact Analysis: Protection of Stratospheric Ozone. Washington, DC.

———. 2000a. Protection of Stratospheric Ozone: Incorporation of Clean Air Act Amendments for Reductions in Class I, Group VI Controlled Substances. Federal Register: November 28, 65:229. 40 CFR Part 82.

———. 2000b. The Kyoto Protocol and President Clinton's Policies to Address Climate Change. http://yosemite.epa.gov/oar/globalwarming.nsf/UniqueKey Lookup/SHSU5BWJUP/$File/wh_c&b.pdf.

———. 2003. Protection of Stratospheric Ozone: Allowance System for Controlling HCFC Production, Import and Export. Federal Register 68:13. Rules and Regulations 2820. 40 CFR Part 82 [FRL-7428–6], RIN 2060 AH67.

———. 2005a. Ozone Depletion Rules and Regulations Fact Sheet-25th Open-ended Working Group Second Extraordinary Meeting of the Parties. June 27–30. http://www.epa.gov/ozone/mbr/MeBr_FactSheet.html.

———. 2005b. Ozone Depletion Rules and Regulation: US Nomination for Methyl Bromide Critical Use Exemptions from the 2006 Phaseout of Methyl Bromide. http://www.epa.gov/ozone/mbr/.

———. 2005c. The Phaseout of Methyl Bromide Ozone Depletion. Rules and Regulations 2005. http://www.epa.gov/ozone/mbr/.

———. 2005d. US Growers Granted Methyl Bromide for 2006 Critical Uses as Annual Requests Decline. Ozone Depletion Rules and Regulations 2005. http://www.epa.gov/ozone/mbr/MeBr_NewsAdvisory.pdf.

US States Say Power Bills Won't Soar on CO2 Plan. 2005. Reuters (November 7). http://today.reuters.com/investing/financeArticle.aspx?type=bondsNews&story ID=URI:urn:newsml:reuters.com:20051107:MTFH82717_2005-11-07_23-16–47_N07555531:1.

Valentine, Mark. 1992. Twelve Days of UNCED: A Follow-Up Report on the Earth Summit. *US Citizens Network Newsletter* (July 2): 3–4.

VanderZwaag, David and Douglas MacKinlay. 1996. Towards a Global Forests Convention: Getting Out of the Woods and Barking Up the Right Tree. In Canadian Council on International Law. *Global Forests and International Environmental Law*. London: Kluwer Law International.

Victor, David A. 2004. *Climate Change: Debating America's Policy Options*. New York: Council on Foreign Relations.

Vidal, John and Paul Brown. 1992. Deadlock in Talks about Aid Cash. *The Guardian* (June 10): 8.

Vig, Norman J. and Michael G. Faure. 2004. Introduction. In *Green Giants? Environmental Politics of the United States and the European Union*, ed. Norman J. Vig and Michael G. Faure. Cambridge, MA: MIT Press.

Voeten, Erik. 2000. Clashes in the Assembly. *International Organization* 54:185–215.

Vogler, John. 1990. The Global Commons. Paper presented at the British International Studies Association Global Environmental Change Group Meeting, February 22, 1991, London.

Vogler, John and Mark F. Imber. 1996. *The Environment and International Relations.* London: Routledge.

Wagner, Lynn. 1998. Problem-Solving and Convergent Bargaining: An Analysis of Negotiation Process and Outcome. Paper presented at the International Studies Association Annual Conference, Minneapolis, MN, March 17–21, 1998.

Walsh, Virginia M. 1997. Ozone Discourses: Science and Politics in Global Environmental Cooperation (review article). *Organization and Environment* 10:318–321.

Wapner, Paul. 2002. The Sovereignty of Nature? Environmental Protection in a Postmodern Age. *International Studies Quarterly* 46:167–187.

Ward, Hugh. 1996. Game Theory and the Politics of Global Warming: The State of Play and Beyond. *Political Studies* 44:850–871.

Webster, P.J., G.J. Holland, J.A. Curry, and H.-R. Chang. Changes in Tropical Cyclone Number, Duration, and Intensity in a Warming Environment. *Science* (September 16): 1844–1846.

Weidner, Helmut. 2002. Environmental Policy and Politics, in Germany. In *Environmental Politics and Policy in Industrialized Countries,* ed. Uday Desai, 149–201. Cambridge, MA: MIT Press.

Weiss, Andrew. 1993. Causal Stories, Scientific Information, and the Ozone Depletion Controversy: Intrusive Scenarios in the Policy Process. In *Controversial Science: From Content to Contention,* ed. Steve Fuller, Thomas Brante, and William Lynch. Albany, NY: State University of New York Press.

Weisskopf, Michael. 1991. U.S. Gets Mixed Reviews on Global Warming Plan; "Action Agenda" Lacks Carbon Dioxide Target. *Washington Post* (February 5): A3.

———. 1992. Bush Was Aloof in Warming Debate; Climate Treaty Offers View of President's Role in Complex Policy. *Washington Post* (October 31): A1.

Welch, Craig. 2000. A Brief History of the Spotted Owl Controversy. *Seattle Times* (August 6): A12.

Wendt, Alexander. 1966. Social Theory of International Politics. Cambridge, UK: Cambridge University Press.

Wettested, Jorgen. 1999. *Designing Effective Environmental Regimes: The Key Conditions.* Cheltenham, UK: Edward Elgar.

Why Toy With the Ozone Shield? 1986. *New York Times* (December 16): A34.

Wilkinson, David. 1999. Unipolarity Without Hegemony. *International Studies Review* 1:141–172.

Willetts, Peter. 1991. Environmental Politics and Regime Theory. Paper presented at the British International Studies Association, Global Environmental Change Group Meeting, February 22, 1991, London.

Woodwell, George M. and Kilaparti Ramakrishna. 1992. Forests, Scapegoats and Global Warming. *New York Times* (February 11): A25.

World Climate Programme. 1988. Developing Policies for Responding to Climate Change. Geneva: World Meteorological Organization.

World Commission on Environment and Development (WCED). 1987. *Our Common Future*. Oxford, UK: Oxford University Press.

World Meteorological Organization. 1985. Report of the International Conference on the Assessment of the Role of Carbon Dioxide and of Other Greenhouse Gases in Climate Variations and Associated Impacts, in Villach, Austria, October 9–15, 1985 (Geneva: WMO Report No. 661).

Young, Oran R. 1991. Political Leadership and Regime Formation: On the Development of Institutions in International Society. *International Organization* 45:281–308.

———. 1993. Perspectives on International Organizations. In *International Environmental Negotiation*, ed. G. Sjöstedt. Newbury Park, CA: Sage Publications.

———. 1994. *International Governance: Protecting the Environment in a Stateless Society*. Ithaca, NY: Cornell University Press.

———. 1998. *Creating Regimes: Arctic Accords and International Governance*. Ithaca, NY: Cornell University Press.

Young, Oran R. and Gail Osherenko. 1993. *Polar Politics: Creating International Environmental Regimes*. Ithaca, NY: Cornell University Press.

Younge, Gary. 2005. U.S. Official Accused of Doctoring Papers Quits. *The Irish Times* (June 13): 11.

Zartman, I. William. 1991. The Structure of Negotiation. In *International Negotiation: Analysis, Approaches, Issues*, ed. Viktor Aleksandrovich Kremenyuk and Guy Faure. San Francisco: Jossey-Bass Publishers.

———. 1993. Lessons for Analysis and Practice. In *International Environmental Negotiation*, ed. G. Sjöstedt. Newbury Park, CA: Sage Publications.

———. 1994. Two's Company and More's a Crowd: The Complexities of Multilateral Negotiation. In *International Multilateral Negotiation: Approaches to the Management of Complexity*, ed. I.W. Zartman. San Francisco: Jossey-Bass Publishers.

Zürn, Michael. 1998. The Rise of International Politics: A Review of Current Research. *World Politics* 50:617–649.

Index